Science and Innovations in Iran

Science and Innovations in Iran
Development, Progress, and Challenges

Edited by
Abdol S. Soofi and Sepehr Ghazinoory

SCIENCE AND INNOVATIONS IN IRAN
Copyright © Abdol S. Soofi and Sepehr Ghazinoory, 2013.

All rights reserved.

First published in 2013 by
PALGRAVE MACMILLAN®
in the United States—a division of St. Martin's Press LLC,
175 Fifth Avenue, New York, NY 10010.

Where this book is distributed in the UK, Europe and the rest of the world, this is by Palgrave Macmillan, a division of Macmillan Publishers Limited, registered in England, company number 785998, of Houndmills, Basingstoke, Hampshire RG21 6XS.

Palgrave Macmillan is the global academic imprint of the above companies and has companies and representatives throughout the world.

Palgrave® and Macmillan® are registered trademarks in the United States, the United Kingdom, Europe and other countries.

ISBN: 978–1–137–03009–2

Library of Congress Cataloging-in-Publication Data is available from the Library of Congress.

A catalogue record of the book is available from the British Library.

Design by Newgen Imaging Systems (P) Ltd., Chennai, India.

First edition: January 2013

10 9 8 7 6 5 4 3 2 1

*Abdol Soofi dedicates this book
to the loving memory of
his son, Rosteen,
April 24, 1975—June 26, 1994*

*Sepehr Ghazinoory dedicates this book,
with thanks for financial support,
to the Center for International Research and Scientific Cooperation*

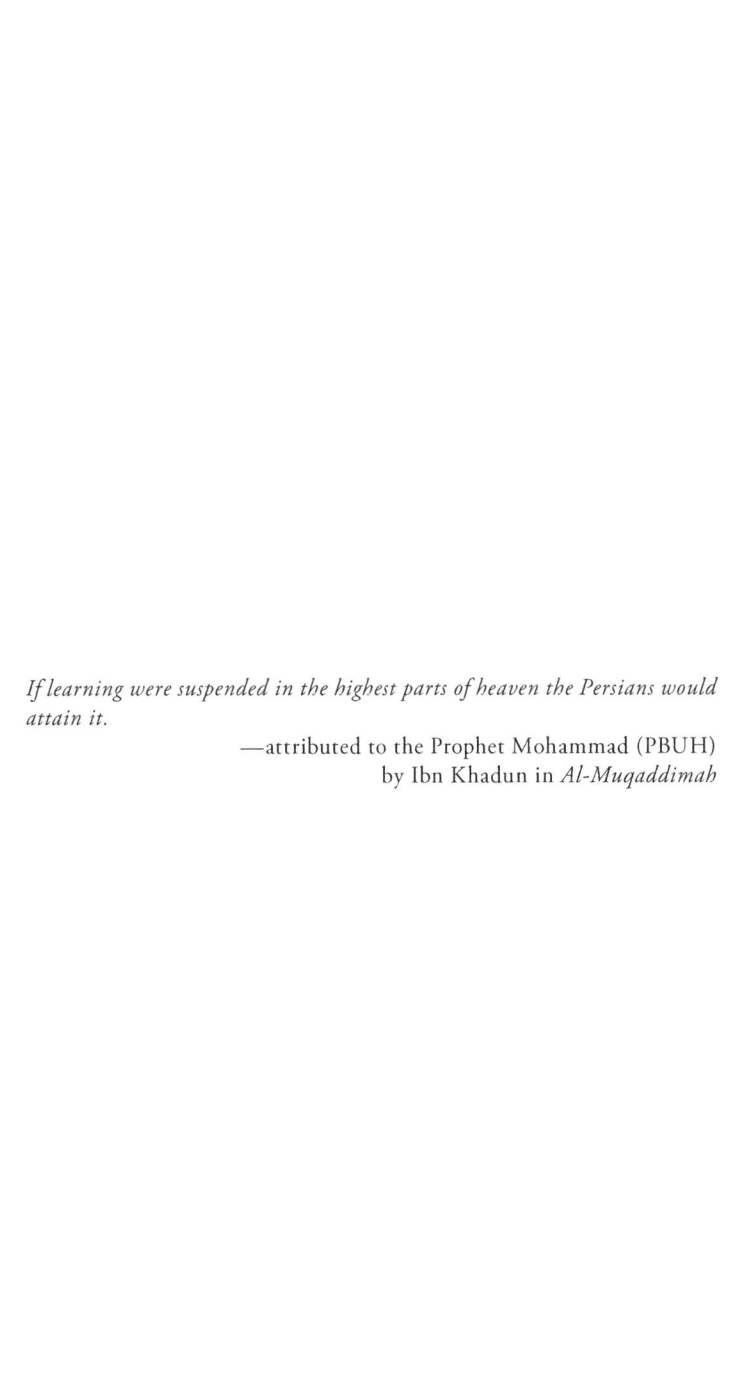

If learning were suspended in the highest parts of heaven the Persians would attain it.
—attributed to the Prophet Mohammad (PBUH)
by Ibn Khadun in *Al-Muqaddimah*

Contents

List of Tables and Figures ix

Editors' Preface xi

Foreword xiii
Hossein Mohammadi Doostdar

1 Introduction 1
  Abdol S. Soofi and Sepehr Ghazinoory

2 The History of Science in Iran from a Physicist's Perspective 15
  Reza Mansouri

3 From Developing a Higher Education System to Moving toward a Knowledge-Based Economy: A Short History of Three Decades of STI Policy in Iran 39
  Mehdi Goodarzi and Soroush Ghazinoori

4 The National Innovation System of Iran: A Functional and Institutional Analysis 57
  Abdol S. Soofi, Sepehr Ghazinoory, and Sanam Farnoodi

5 Information and Communication Technology: Between a Rock and a Hard Place of Domestic and International Pressures 87
  Sepehr Ghazinoory and Reza Jamali

6 Nanotechnology: New Horizons, Approaches, and Challenges 115
  Fatemeh Salehi Yazdi and Mohammad Ali Bahreini Zarj

7 Biotechnology in Iran: A Study of the Structure and Functions of the Technology Innovation System 139
  Tahereh Miremadi

8	Nuclear Technology: Progress in the Midst of Severe Sanctions *Behzad Soltani and Marzieh Shaverdi*	159
9	Iran's Aerospace Technology *Parviz Tarikhi, Mohammad Abbassi, and Maryam Ashrafi*	185
10	The Automotive Industry: New Trends, Approaches, and Challenges *Manochehr Manteghi*	217
11	Conclusion *Sepehr Ghazinoory and Abdol S. Soofi*	245

List of Contributors	249
Name Index	251
Subject Index	255

Tables and Figures

Tables

2.1	Indicators of Educational-Scientific Development, 1979–2012	33
3.1	Key Indicators of Research and Technology Development in Iran, 2005–2009	48
3.2	Policy Documents on Science, Research, and Technology Formulated after the Revolution in Iran	51
4.1	Entrepreneurial Attitudes and Perceptions: Select Countries	61
4.2	Participation of Individuals in Entrepreneurial Activity	64
4.3	Education Statistics	71
5.1	Trends of e-government in Government Agencies	101
5.2	Indicators of Communication and Information Technology Development in Iran	107
6.1	National Priorities in Nanotechnology	119
6.2	Technology-Market Services Corridor (TMSC) Services	124
6.3	Pilot Projects Run by the INIC and Their Current Status	131
7.1	Timetable of Major Events in the Course of SCRTIS Building	151
8.1	Nuclear Technology Achievements in Iran	175
10.1	Iranian Automobile Firms and Their Output, 2010	226

Figures

1.1	Coordination Failures and Government Remediation	6
2.1	History of Science Institutes in Iran	22
4.1	Governance of Science, Technology, and Innovation in Iran	70

Editors' Preface

Over the last two decades, Iran has gone through a major industrial transformation in spite of "abnormal" obstacles in the path of the country's development. All indicators point to the industrial "takeoff" of the economy. However, little is known in the English-speaking world about these fundamental developments and the challenges the policy planners faced and continue to face in the development processes.

This book is an attempt to fill this knowledge gap by critically examining, first, the Iranian government's science, technology, and industrial policies that support the industrialization and national defense of the country, and, second, by reviewing the innovation activities of Iranian enterprises.

Specifically, the book will, for the first time, present in a single volume an overview of the organizations, policies, conduct, and performance of those government organizations that have the task of developing traditional as well as emerging technologies in the country. In short, the book is a collection of chapters that allows us to gain an understanding of the structure and dynamics of the science and technology policies of the Iranian government. Furthermore, the chapters will critically examine the major challenges that the country faces in developing traditional and emerging technologies.

The Iranian government plays a pivotal role in development of technologies and the production of aggregate output in the country. With the state taking a prominent role in technological development in Iran, and toward the characteristic uniformity of the proposed chapters in the book, the editors invited a group of authors to contribute chapters related to a broad outline of the proposed book, and have the honor of having had their invitation accepted. Most of the authors are in a position of leadership in organizations devoted to the creation of knowledge bases in relevant technologies, and all of them have intimate knowledge of the activities of the institutions and agencies with which they are associated. This volume is the product of the collective efforts of all of the contributors to this book. Accordingly, the views expressed in the following chapters are those of the authors of the chapters and not of the editors of the book.

We are grateful to Ms. Evelyn Martens for her editorial assistance in reviewing the chapters.

Foreword

The important role that Iran is taking in international scientific cooperation and the pivotal role it is playing in learning, gaining experience, and contributing to two-way transfers of advanced science and technology between itself and the international community is not a phenomenon that is familiar to everyone. In order to establish a powerful, effective presence on this stage and achieve success in attaining this goal, all Iranian researchers, scientists, and managers as well as their international counterparts must be aware of each other's capabilities in science and technology. This awareness is a precondition for the expansion of the initiation of and the support of ongoing scientific and technical collaborations and interactions between Iranian researchers and their international colleagues.

The Center for International Scientific Studies and Cooperation [CISSC] has been established to support scientific activities, promote the development of technology, and expand the relationship between Iranian universities and research centers on the one hand, and international scientific centers on the other. Accordingly, the Center, in order to create a suitable framework for promoting efficacy in the interactions between Iranian scholars and researchers and their international collaborators, provides financial assistance as well as technical and moral support to researchers and scholars in Iran.

The present book aims to introduce the achievements of Iranian scientists, researchers, and industrialists who are active in the science and technology fields to the international community. By supporting this project, which is the result of a set of timely, insightful analyses concerning the latest scientific and technological developments in Iran that make up the chapters of the book, the Center believes it is promoting further research and development activities in Iran, and by doing so, it has implemented its mission in this regard.

HOSSEIN MOHAMMADI DOOSTDAR
Head, the Center for International Research
and Scientific Cooperation

CHAPTER 1

Introduction

Abdol S. Soofi and Sepehr Ghazinoory

Introduction

Developing economies often achieve industrial development and catch up technologically in normal development circumstances through the acquisition of foreign technologies by means of mechanisms such as purchasing equipment from abroad, licensing, subcontracting, engaging in foreign direct investment, hiring foreign experts, participating in joint ventures and the formation of strategic alliances, manufacturing original equipment, and purchasing foreign enterprises outright. However, under the conditions of international sanctions and limited interactions with technologically advanced countries, many of these approaches are not readily available for countries facing economic sanctions. Iran, of course, is confronting such sanctions, and cannot use many of the stated mechanisms listed above to transform its industry. Therefore, the country must rely on its own physical and manpower resources as well as the ingenuity of its citizens for technological advancement.

Moreover, it is widely known that even under normal conditions of development, many emerging economies, even though they may successfully adopt and imitate foreign technologies, face formidable barriers and challenges in moving beyond imitation and becoming innovators. Iran is no exception in this regard. Accordingly, Iran faces two sets of challenges in resolving its latecomer status in the technological arena.

This book is an attempt to gain an understanding of how Iran is coping with these two sets of constraints on the path to its industrial transformation, how successful it has been, and what challenges it faces in achieving its developmental objectives.

In this introductory chapter, we will examine the broad theoretical framework of the industrial transition of emerging economies as a guide for evaluating the industrialization processes in Iran. This theoretical discussion, along with Iran's experiences in technological-industrial development, will allow the reader to gain an understanding of whether the country has followed and is following the guidelines, which are based on the experiences of the newly industrialized economies (NIE) of East and Southeast Asia. Moreover, we will delineate the current state of technological development in Iran, and based on a review of the chapters in this volume, discuss the country's efforts in overcoming the technological challenges it faces.

Coordination Failures: Market and Government

This study, to a large extent, is concerned with the industrial development policies and practices of different administrations in Iran since the Iranian Revolution of 1979, with particular emphasis on the 1990–2010 period. These policies are tantamount to the coordinating activities of the government in the country. Since private property is the dominant form of production relationship in the country, and since most exchanges are mediated by means of markets, the prominent role of the government in the process of the industrialization of the economy implies the failure of the markets to coordinate economic activities. However, coordination failures, in the sense defined below, are so pervasive in the Iranian economy that one could reasonably attribute the failures to both the government and the market. Therefore, defining economic coordination is of paramount importance.

Economic coordination refers to a "pleasing" arrangement of economic activities. For example, Klein (1997) states: "Coordination ... is best understood as something we hope to achieve in our interaction with others." Viewing economies from this perspective, one could consider an economy with full-employment output successful; therefore, a full-employment level of output is deemed to be the result of a coordinated economic system. Or, based on this definition, one would conclude that coordination is a failure (market, government, or both) in a depressed economy, where markets for labor, capital, and goods do not clear, and excess supplies emerge.

The second definition of coordination is a mutual meshing of activities, and in the economic context, one would consider market clearance as a coordination activity, or market disequilibrium as a coordination failure.

Traditionally there were two opposing views on the role of the government in developmental processes: the market-friendly view and the developmental-state view. The market-friendly view holds that most, if not all, coordination activities in an economy are best performed by the market mechanism. Government intervention in the marketplace, according to this view, should take place only in cases of market failure. Furthermore, the market-friendly view states that even under the best of circumstances where the government successfully coordinates a particular project, it blocks market-coordination over the long run, and hence makes sustainable coordination impossible (Matsuyama 1996). The developmental-state view, in contrast, perceives market failures in developing economies to be pervasive and looks to government to remedy these failures (Aoki, Kim, and Okuno-Fujiwara 1996).

A third view, as presented by Aoki and colleagues (1996), departs from the neoclassical economic theory, which predicts that competitive markets tend to allocate resources optimally, and argues that governments can "enhance" market performance, thus implicitly rejecting the market as the optimal resource allocation mechanism. This is called the "market-enhancing" role of government (Aoki, Kim, and Okuno-Fujiwara 1996).

In the literature dealing with the "Asian Miracles," where the term "miracle" refers to the rapid industrialization of the East Asian economies of South Korea, Taiwan, Hong Kong, and Singapore (NIEs), the same controversy regarding the preeminence of the state versus the market in the developmental process emerged also. The proponents of the primacy of the state in the industrial development of these economies include Kim and Dahlman (1992), and Amsden (1989), while those who believe that after the initial government-led coordination, firms and by implication, the markets take control of coordination and play a more important role in the coordination process include writers such as Hobday (1995), Hobday, Cawson, and Kim (2001), and many other scholars.

The historical experiences of many newly industrialized countries of East and Southeast Asia, as well as the important role of the Iranian government in the country's economy, as documented by the chapters in this book, give substantial support to the "market-enhancing" view of the role of the state in economic development, and we uphold this view in the present volume.

Setting the macroview of debate on the relative importance of state versus market coordination aside, we find the examination of the patterns of development of successful industrial enterprises in East Asia during the last three decades insightful and illuminating. We examine two important contributions in this field of inquiry: Hobday and colleagues (2001) and Lee and Lim (2001). These authors critically reviewed the transformation of several local firms in South Korea and Taiwan, and showed how these enterprises, as latecomers, were able to acquire technical and marketing competencies and become globally competitive.

Hobday (1995) classifies firms as "leader" "follower," or "latecomer." Technology leaders are innovative firms that operate in markets with an advanced technological infrastructure and develop new products and processes. Follower firms are directly connected to the advanced markets; however, their technology lags behind that of the leaders. Nevertheless, they have the opportunity to learn from the existing technologies, improve upon them, and even become leaders in their own right.

The leader-follower classification of firms characterizes firms that operate in the national system of innovation in advanced industrial economies. A different class of firms that is often encountered in the developing economies is called the "latecomer" firm. Latecomer firms face two formidable technological and market-access challenges they must overcome if they wish to succeed globally.

Being located in developing countries, the latecomer firms are isolated from the main centers of technological development, have little or no access to distinguished institutions of higher learning, operate in business environments with poorly developed technological infrastructure, are behind in engineering skills, and cannot initiate meaningful, effective research and development (R&D) projects. In most developing economies, the marketing problems faced by these latecomer firms include small domestic markets, which cannot provide the opportunities for the firms to achieve economies of scale, and high barriers to entry into the foreign markets that enjoy high demand (Hobday 1995). To be globally competitive, firms in the developing countries must overcome these technological and marketing barriers.

According to Lee and Lim (2001), not all latecomers follow the same path of evolution to follower and leader status, however. Some latecomers skip the developmental processes, perhaps creating their own pattern of evolution that is different from that of their forerunners. Studies have shown that not all industries within a newly industrialized economy or the same industry in different countries has followed the same path. It is even possible to observe the heterogeneous development of industries within a country.

Furthermore, analysis of the industrial development of successful latecomer firms has indicated that, in general, these firms in the emerging economies go through three phases, starting with original equipment manufacturing (OEM), moving to own design manufacturing (ODM), and from there entering into own brand manufacturing (OBM).

OEM refers to subcontracting where a foreign buyer (often a transnational company) provides the detailed technical specifications to the latecomer firm for manufacturing the product. In an ODM system, the latecomer both designs and manufactures the product, and then sells it as OEM. Finally, in an OBM system the latecomer has developed the capability of designing, manufacturing, and marketing to challenge the leader and compete globally (Hobday 2001).

These broad discussions relating to the relative roles of the state, markets, and firms in the developmental processes in NIEs over the last five decades provide a framework for us to examine and assess the role of the state, markets, and enterprises in the evolution of the Iranian economy.

First, it should be noted that three classes of enterprises exist in Iran: state-owned enterprises (SOE), cooperatives, and private enterprises. In the manufacturing sector, the SOEs play the most prominent role among the three. Accordingly, the state in Iran must adopt those policies that remedy both the coordination failures of private and cooperative firms, through direct and indirect measures, as well as address the coordination failures of SOEs, which have the responsibility for managing the government-owned firms and for acting as the governing bodies of enterprises.

Second, since the Iranian government is deeply involved in the design and implementation of industrial as well as science and technology policies for industrial development, we need to define industrial development, industrial policy, and science and technology policy in order to have a better understanding of the analyses that appear below and in the succeeding chapters of this book.

First, "industrial development" is defined as "a process of acquiring technological capabilities..." (Kim and Dahlman 1992, 437). Next, Kim and Dahlman (1992, 437) define science and technology policy as "a set of instruments the government uses in promoting and managing the process and direction of acquiring such capabilities." Accordingly, we consider technology policy as a set of *direct* and *indirect* governmental policies that affect scientific and technological development. The direct policies define the direction and pace of the supply side of technological development by creating and strengthening technological capabilities. This is commonly known as science and technology policy. The indirect policies may be divided into two subcategories. The first subcategory involves a set of policies that affect the demand side of technological development by creating the need for technological capabilities through financial and tax incentives to firms. The second subcategory consists of a set of policies that facilitate coordination (in the sense defined above) by creating conditions that ensure the smooth functioning of the market for technologies. The indirect policies are referred to as industrial policy.

The Role of State and Market in the Successful Industrial Transformation of Enterprises

Figure 1.1 is the flow chart of interactions between firms and the state in the process of industrialization. Coordination failures by the market

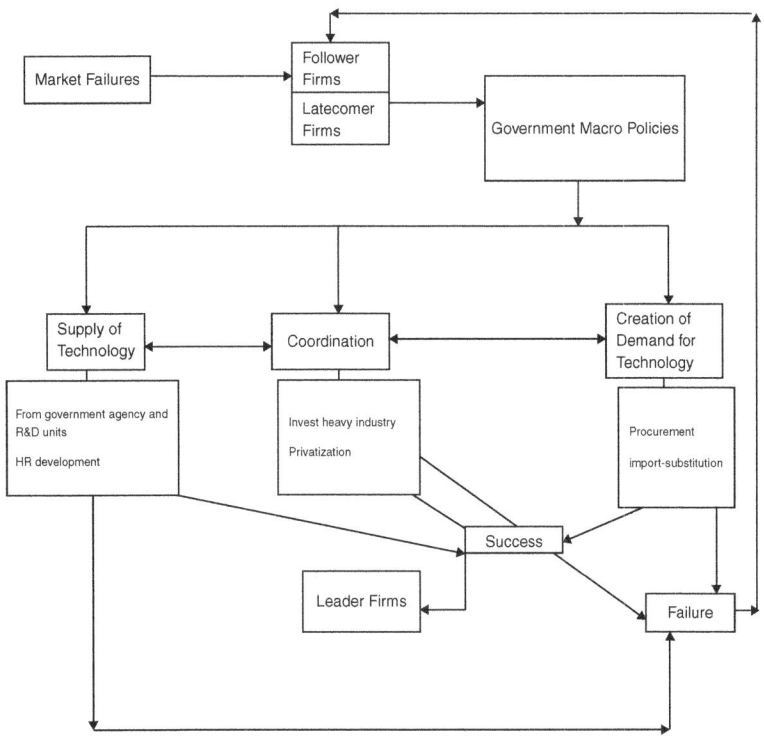

Figure 1.1   Coordination Failures and Government Remediation.

lead to the latecomer/follower status of firms. The government's attempt to remedy market failures by taking policy measures to supply technologies, induce private or state-owned enterprises to demand technologies, and to adopt policies to clear the markets for technologies can lead to either success or failure. The government's failure to achieve its objectives leads to the status quo for latecomer/follower firms. In contrast, government success leads to leadership status of the latecomer/follower firms, which is tantamount to industrial transformation and progress.

What are the specifics of supply-side, demand-side, and coordination policies? What specific policy measures could governments use to remedy market failures, which have resulted in the formation of latecomer/follower firms? Even though many policies have been adopted by governments of the NIEs over the past 40 years, we state a sample of measures in each category that have been used in the past, particularly in South Korea. It should be noted that by no means is the list exhaustive.

(i) Policies to induce demand for technologies by latecomer firms

Measures to induce demand for technology include government procurement of products and services; import-substitution policies; government policies to reduce market power and enhance competition; export-promotion policies; industrial policies to strengthen target industries (e.g., imposing quotas, import licensing, domestic content requirements, offering tax incentives, preferential financing, loan guarantees, and R&D subsidies in the form of government-corporate research for those who produce the designated products); and protection of intellectual property rights. Moreover, establishment of a central agency to coordinate the technological activities of all ministries of the government would have both supply-and-demand effects in building technological capability.

(ii) Policies to supply technologies to the private sector

This set of policies can be divided into direct and indirect policy instruments. Direct policy instruments include the creation of a central government agency, such as a Ministry of Science and Technology, to coordinate the technological activities of other ministries, and establish R&D organizations. Indirect policies consist of investment in human resources; import-substitution; restricting technology transfer and foreign direct investment (FDI); promotion of turnkey plants and the import of capital goods in the early stage of industrialization; lifting of the restrictions on FDI and foreign technology transfers at a later stage of the development; export promotion; and import of foreign capital. Moreover, governments have established highly specialized research institutes and firms as spin-offs from earlier public research institutes to develop in-depth capabilities in targeted industries.

(iii) Coordination policies
Coordination policies that are used include the creation and promotion of the heavy industries (steel, chemicals, and metal cutting machinery, shipping, other means of transportation, and infrastructure); policies to achieve economies of scale, like *Chaebols* (large conglomerates) in South Korea; development of "strategic industries"; privatization of SOEs; allocation of credit to select industries; a preferential exchange rate for use by large enterprises; state guarantee of foreign loans borrowed by *Chaebols*, for the import of turnkey factories for import-substitution industries, and supplying low-interest-rate loans to firms with losses due to foreign currency exposure; further licensing of successful *Chaebols*, the penalizing of poor performers by not supporting them and allowing them to fail; and the granting of licenses in lucrative industries to those enterprises that are willing to invest in riskier sectors. Other coordination (linkage) policies used in South Korea include tariff exemptions on capital goods; low interest-rate loans by foreign suppliers; establishment of a technical transfer center that informs domestic firms about available technologies abroad; establishment of public R&D institutes to act as transfer agents by undertaking joint R&D with private firms, enabling them to gain adequate knowledge about available foreign technologies; provision of information about the suppliers of technologies; establishment of technical extension services for small and medium-size firms; and the creation of technical information centers.

It should be noted that, at least according to the South Korean experience, in the absence of demand for technological change, the direct instruments were ineffective in increasing the supply of technology, while the indirect instruments were more effective. The ineffectiveness of direct instruments such as R&D institutes occurred because there was no demand for the type of expertise the institutes could offer. In the meantime, these institutes lacked the manufacturing know-how that was in great demand by the industries (Kim and Dahlamn 1992).

Additionally, these policies have different effects on technological capacity-building, depending on the stage of development of the industries. According to the evolutionary development of technology, the industries in advanced industrial economies go through three stages: emergence (product innovations involving high risk and high commercial failure), consolidation (process technology changes rapidly, lowering the cost of production), and maturity (further innovation becomes difficult, which leads to the loss of competitive advantage) (Dosi 1982).

Contrary to the evolutionary development of technology in the developed economies, and based on the industrial experiences of the NIEs, industries in the developing countries start at the mature stage, proceed to the consolidation stage, and finally enter the emergence stage. The patterns of technology policy and development in the NIEs, as delineated above, could provide important lessons for Iranian science and technology policy planners.

The chapters in this volume should be read within the framework of the stylized scheme above. As stated above, the manufacturing enterprises in Iran fall into three categories: state-owned, cooperatives, and private. Of course, whether the SOEs are managed as for-profit or not-for-profit enterprises makes a fundamental difference in whether firms in Iran follow the same evolutionary pattern of development of enterprises in the NIEs. The problem of corporate governance is a fundamental issue in industrialization processes; however, discussion of that important subject is outside of the scope of this volume and requires a separate study.

Most of the chapters in this volume deal with various technologies. Some of these technologies, such as nanotechnology, stem cell, and biotechnology, are in their infancy, and have not found widespread use in the production processes. Other technologies have found application in the industries that are in the mature stage, for example, automobile, aerospace, nuclear, and information.

* * *

In chapter 2, Reza Mansouri gives an insightful historical account of the development of science from ancient Persia to modern Iran, arguing that the notions of science and scientists as defined in the West have different connotations in Iran. Furthermore, Mansouri argues that due to the clerical antagonistic view toward the natural sciences, scientific theories and methods as known in the West did not progress in Iran and the other Islamic countries, until the establishment of the Islamic Republic (IR) of Iran in 1979, when Iranian nationalism created suitable conditions for what he terms a "scientific awakening." In spite of recent gains, Mansouri asserts that the country is facing severe impediments as far as the acceptance of modernity is concerned.

In chapter 3, Mehdi Goodarzi and Soroush Ghazinoori trace the historical development of education, the sciences, and research in Iran, and again conclude that much of the progress in these areas took place after the establishment of the Islamic government in Iran.

In chapter 4, Abdol Soofi, Sepehr Ghazinoory, and Sanam Farnoodi apply the functional approach to the study of the innovation systems of Iran. They find that the Iranian system, in spite of major progress over the last two decades, faces the challenge of converting knowledge to wealth at the industry level. Using the historical experiences of the West and Japan, they argue that converting knowledge to inventions, inventions to innovation, and innovation to wealth requires the endogenous development of industries and markets in a country and beyond, a process that takes decades to complete. Moreover, the authors explain that the Western economic sanctions against Iran have created a bittersweet condition for the country. On the one hand, the sanctions have reduced, if not completely eliminated, the transfer of technology from abroad. At the same time, they have stimulated efforts in the country to rely on its human resources and domestic capabilities to develop science and technology independent of imported technology.

Chapter 5 is about communication and information technology. Sepehr Ghazinoory and Reza Jamali critically examine the information and communications technology (ICT) sector of the economy and conclude that in spite of major technological gains in the industry over the last decade, certain constraints have created impediments for further growth of the sector. They state that the culture of software piracy and the violation of intellectual property rights in Iran are prevalent. This tendency is attributed to Islamic jurisprudence, which does not accept the doctrine of the universality of the institution of private property, and upholds the notion of common property in certain cases. Moreover, the authors cite national security and moral concerns on the part of Iranian government authorities concerning potential abuses of the Internet by foreign powers to wage cultural wars and create political instability in Iran as having contributed to the government's policy of severe filtering of the Internet and other censorship measures. Accordingly, they believe that as a result of government suspicion of ICT applications in Iran, this technological sector, contrary to the other technological sectors, does not receive wholehearted support from Iranian leaders. Furthermore, they state that due to underinvestment and limited markets for outputs of the sector, the ICT industry faces diseconomies of scale and inordinate costs of access to the Internet.

Based on these considerations, the ICT sector in Iran faces both international and domestic countervailing pressures for more rapid developmental processes. Perhaps because of these opposing forces, a myriad of political, religious, cultural, security, and military establishments are engaged in ICT decision-making in the country.

Chapter 6 examines nanotechnology development in Iran. Fatemeh Yazdi and Mohammad Ali Bahreini Zarj state that unlike with previous experiences of technology development in the country, Iran decided to develop nanotechnology at the *emergence state* of the technology. The country chose to develop the technology selectively, not only to develop technological capabilities in this field, but also as a benchmark for the development of other technologies. A study of Iran's nanotechnology program that identifies its successful experiences and its mistakes and policy failures, holds valuable lessons for policy makers and policy researchers.

Yazdi and Bahreini Zarj examine several important policy lessons that technology planners learned from developing strategy and implementing it. These lessons include the importance of promoting the technology to create awareness among experts and the public, the significance of developing proper goals, and the importance of relying on the country's own resources and developing its self-sufficiency in the face of international sanctions.

The less successful aspect of the nanotechnology initiative, according to Yazdi and Bahreini Zarj, includes an inadequate emphasis on revenue-generating aspects of the program through the commercialization of R&D results.

In chapter 7, Tahereh Miremadi undertakes a case study of the Royan Institute through a structural mapping of the actors, networks, and institutions that are part of the technological innovation system (TIS) of stem cell research in Iran. Furthermore, she evaluates the functional pattern of the system. In her functional analysis, she discovers a dynamic system below the surface of this static picture created by a major player, that is, the Royan Institute.

The Institute began its self-proclaimed mission by contributing to the legitimization of the technology project. As a result, public expectations grew increasingly higher after the country achieved the landmark scientific events of registering the first and only Iranian embryonic stem cell line and additional successful experiments. The third major step was the mobilization of resources and the acquiring of a governmental budget line. The supportive institutional environment and the Royan Institute's contributions and networking, which extend far beyond the usual entrepreneurship knowledge, are the facilitating mechanisms in the process of institution building and networking, and they reveal the Royan's role as the driving motor of system building. The Royan Institute has ended up creating significant scientific and intellectual capital in TIS, influencing some of the system's functions. However, the momentum that the Royan's contributions have single-handedly created has failed to encourage some other crucial functions such as the entrepreneurial experimentation function, market formation, and development of positive externalities.

Chapter 8 examines nuclear technology development in the country. Behzad Soltani and Marzieh Shaverdi critically review the history of the development of the industry since the 1950s, and describe the outstanding achievements and challenges that the industry faces. In addition to listing the technological advances of the industry over the past six years, they give a detailed account of obstacles in the path of further development of the industry. They further argue that in spite of severe international sanctions, Iranian researchers and technologists have made notable progress in many areas of nuclear technology. They state that the development of these technologies was made possible by the cooperation of many entities and with the strong support and guidance of the Supreme Leader, Ayatollah Khamenei, the executive branch under the leadership of President Mahmud Ahmadinejad, and the legislative branch of the government.

Furthermore, the authors state that the relentless pursuit of nuclear technology in the face of the severest international sanctions against the country has assumed a dimension that goes beyond pure economic calculations and a desire to acquire the technology. They believe that the pursuit of acquiring a mastery of nuclear technology has become a symbol of national self-determination, independence, and pride. They rightfully assert that the IR of Iran does not demand extralegal rights for the development of nuclear technology in the country, is fully committed to the principles of the United Nations Nuclear Non-Proliferation Treaty (NPT), and insists on protecting its rights as a member state to pursue the development of nuclear technology for peaceful purposes.

Chapter 9 deals with technological development in the Iranian aerospace industry. Parviz Tarikhi, Mohammad Abbassi, and Maryam Ashrafi describe the history, current status, and future plans of Iran's aerospace program. They argue that Iran's efforts to develop aerospace technology for civil and peaceful purposes began two decades ago, and they attribute almost all the technological successes of this sector to indigenous potential in terms of human resources and available expertise and experience, along with international cooperation and the exchange of knowledge. They state that there are considerable civilian entities involved in space-related development and production in Iran. Furthermore, they argue that in recent years, Iran has developed strategic plans for the development of an aerospace industry that is internationally competitive. Accordingly, the Iranian government has supported the aerospace research activities of research centers and universities throughout the country. They note that such achievements require a high degree of expertise, ability, and comprehensive knowledge about the subject, and that the supportive vision and attitudes of Iran's leaders have contributed to the pace, progress, and developmental objectives of the nation's aerospace program.

Chapter 10 discusses the technological development of the automotive industry. Manochehr Manteghi critically examines the history and evolution of the industry from its humble beginnings as an industry consisting of a firm that assembled foreign-manufactured automobile parts in the late 1950s. Manatghi asserts that the socioeconomic and political events of the last half-century in Iran had a drastic retarding effect on the nature and rate of growth of the Iranian automotive industry. However, in spite of these setbacks, he argues that to a certain extent, acquiring the technological knowledge, in cooperation with certain prominent world automakers, is underway, and that these new technologies have been incorporated into domestically produced vehicles. Nevertheless, some lingering problems still persist on the path to full technological development of the industry. He cites examples such as protective tariffs, which have granted the industry—in fact only two leading firms—monopoly status. Moreover, because of the traditional commercial (in contrast to industrial and innovative) views of many auto executives, their inability to effect innovation in their firms, and limited government involvement in promoting the industry's investment in R&D, the development of the automotive industry is less than that expected by world standards. This has led to inadequate growth of the industry in global competition.

The editors provide some concluding remarks in the last chapter. They observe that in spite of many political, economic, historical, and cultural constraints, the defense industry in Iran has been more innovative than the civilian industries in the country. They attribute the difference in performance to the structure of the industry and the management role of the Iranian armed forces, which are under direct control of the Supreme Leader, and which insulate technological policy making from the volatile political environment of the country.

References

Amsden, A. 1989. *Asia's Next Giant: South Korea and Late Industrialization.* New York: Oxford University Press.

Aoki, M., H. K. Kim, and M. Okuno-Fujiwara. 1996. *The Role of Government in East Asian Economic Development: Comparative Institutional Analysis.* Oxford: Oxford University Press.

Dosi, G. 1982. "Technological Paradigms and Technological Trajectories." *Research Policy* 11: 147–162.

Hobay, M. 1995. "East Asian Latecomer Firms: Learning the Technology of Electronics." *World Development* 23: 1171–1193.

Hobday, M., A. Cawson, and S. Ran Kim. 2001. "Governance of Technology in the Electronics Industries of East and South-East Asia." *Technovation* 21: 209–226.

Kim, L., and C. J. Dahlman. 1992. "Technology Policy for Industrialization: An Integrative Framework and Korea's Experience." *Research Policy* 21: 437–452.

Klein, D. B. 1997. "Planning and the Two Coordinations, with Illustration in Urbantransit." *Constitutional Political Economy* 8: 319–335.

Matsuyama, K. 1996. "Economic Development As Coordination Problems." In *The Role of Government in East Asian Economic Development: Comparative Institutional Analysis,* edited by M. Aoki, H. K. Kim, and M. Okuno-Fujiwara, 134–160. Oxford: Oxford University Press.

CHAPTER 2

The History of Science in Iran from a Physicist's Perspective

Reza Mansouri

Introduction

I am not and do not pretend to be a historian of science. However, I am a scientist who knows about the state of the sciences in Iran at the time when I was an elementary and high school student in Tehran in the 1950s. I completed my university education in Vienna, a city in which those traditions such as Heinrich Hertz, Ernst Mach, Erwin Schrödinger, and the Vienna School of Philosophy were still alive. Since the Islamic Revolution, I have continued to play a role in the evolution of science in Iran. I have acquired a keen understanding of the relationship between Iranian thought and the different meanings of science. When I was in elementary school, our living conditions and the facilities to which we had access in Tehran were not much different from the conditions that prevailed eight hundred years ago: We had no electricity, no televisions, no refrigerators. I still recall the dusty streets after the first public city bus lines were established in the neighborhood where we lived.

After receiving my high school diploma, I thought about studying astronomy abroad. In all the libraries and bookstores in Tehran, one could hardly find more than ten books on astronomy.

I busied myself with books by Abu Rayhan Birooni's At-tafhim, Abdulrahim Soofi's Sovar –ol-kavakeb, and Ghathan Ibn Marvazi's Kayhan Shenakhat. Finally, when the owner of the bindery where I worked at that time, an educated middle-class man with a habit of reading, learned that I was interested in astronomy, he suggested that the best place to study the sciences (*elm*) was in Najaf.[1] He meant the Najaf seminaries, which were considered more prestigious than those in Qom. Understanding why the middle class, remarkably literate in Iran in 1965, had such a perception of the sciences is not easy.

I am sure that for many of our university students, this reality belongs to the past, belongs to hundreds of years ago, but it was a part of my scientific life. It should be noted that our scientific output under the best of conditions was not more than one hundred articles in international journals in those years. Currently, the number of scientific publications from Iran has reached approximately twenty thousand annually.

Based on this description, and based on the historical presence of great thinkers such as Birooni, Soofi, Avicenna, Razi, and others, how can one understand the history of science in Iran? This is the question that Iranians have asked themselves for over one hundred years: Why is Iran still a developing country and has it not been successful in overcoming the obstacle of backwardness? This is a discussion that is very difficult to separate from similar developments in other Islamic countries (Hoodboy 1992).

In this chapter, based on the concept of modern science, as a complex social phenomenon, I have tried to analyze our own history, and hope to play a role in these continuing discussions. My heterodox views about science and technology in Iran and its relationship with the evolution of the modern sciences and technology over the last two hundred years are rooted in my interest in the history of the sciences as well as my engagement in activities focused on the development of science in Iran following the Islamic Revolution. This perspective deeply diverges from the common arguments concerning the role of colonialism or Islamic rationalism that appear in many studies concerning science in the Islamic countries (see, for example, Sardar 2007 and Serageldin 2007).

Discussing the state of science and technology in the Islamic countries, including Iran, is complicated, especially in terms of communication with people in the industrial world. The use of terms such as "science" and "technology" usually leads to an implicit misunderstanding that makes dialogue on this subject very complex. The reality is that terms such as "research, science, and scientist" cannot be easily translated or directly mapped into Farsi. For example, in Farsi, *elm* is translated as "science"; however, in reality this classical term does not represent the complex social process that the modern world would recognize as science. In Persian, a person possessing or being engaged in *elm*/science is called *ahl-e-elm* or *aalem*. *Aalem* is a widely used term in Moslem culture. It is, however, a common understanding in Iran that one who is *ahl e elm* is a man who has spent years in a seminary, and whose studies were guided by a well-known scholar of Islamic religion. A modern scientist would never be called *ahl-e-elm*. The Persian word *daneshmand*, a term more than a thousand years old, is sometimes taken as equivalent to scientist, with the latter term coined around the mid-nineteenth century in England. However, the word *daneshmand* hardly suits the context of the English term scientist. *Daneshmand* is a general term for scholar, or natural philosopher, but not scientist. I have coined the term *daneshgar* to be used as an exact translation of the term scientist: a person who does science or has the profession of science. But this new term, being derived from the term *danesh not elm*, is not yet widely accepted, and many are hesitant to use it.

The Main Components of Modern Science and Technology

Without an adequate knowledge of the phenomena of modern science and technology, and the features distinguishing them from what is called science in Islam, it is difficult to conduct a historical analysis of science in Iran. The historians of the Islamic sciences, independent of the case studies, are usually divided into two groups in expressing general ideas about the subject: Either they emphasize the importance of science in the Islamic epoch for the historical development of the sciences for all humanity, or they discount its importance in successive scientific developments. Both groups are influenced by romantic interpretations of either the Christian West or the Moslem East. From my perspective, the development of Islamic science is part of the evolution of human thought, which is notable, and the nature of its growth and decay for humanity, and particularly for Moslems, is very informative. In current discussions relating to the development of the sciences in Iran since the Islamic revolution, inadequate knowledge of the evolution, either during the Islamic Republic or before, has caused many misunderstandings that slow down the pace of our scientific development.

Modern Science as a Social Phenomenon

The progress of the modern sciences depends on considering science as a social phenomenon. Modern science is a process in which scientists, scientific institutions, government, and society play a role and determine its dynamics. Lack of attention to any of these components in any society is a sign that the society is still in a premodern era. Two concepts from the scientific community and scientific discourse play a key role in the progress of the modern sciences. The verification of scientific work, scientific thought, and the empirical results of science are determined by the scientific community, and scientists earn their scientific credibility from that community, not from political and social institutions. Scientific discourse as employed in conferences, the publication of articles, and scientific reports helps scientists to earn credibility within the scientific community, and simultaneously determines the dynamic process of the modern sciences (Mansouri 2007b; Shi 2001).

The Relationship between Human Beings and Nature in the Modern Sciences

In the modern sciences, nature is a reality, which scientists aim to understand by modeling, that is, by identifying the concepts and articulating the relationships among them. Accordingly, the scientist is no longer interested in understanding the nature of the concepts of time and space, but considers these concepts as a construct of human imagination. By establishing relationships among them, the scientist is then in a position to predict and measure in obtaining answers from nature. Modeling and constructing concepts that are related to the models establish a relationship between man and nature, which is different from that conceived by scholars before the modern era, including in the Islamic Golden Age. Moslem scholars were also seeking to understand the nature of concepts like time and space. This type of modeling has empowered humans to predict phenomena in nature, as well as has allowed for the creation of new materials and the building of complex systems. In this approach to modeling, reductionism has a central role, to the extent that some scientists consider it the only way to understand nature, even though some biologists disagree with this view.

Homonymity in Scientific Concepts

To begin, I would like to distinguish among four concepts, all of which are referred to as science in Iran (Mansouri 2007b):

> The traditional theology, which is taught in religion seminaries (*Hozeh elmiyeh*); this category of knowledge is based on book (texts) and is called science (*elm*) in Iran and all Moslem countries.

The traditional natural sciences that existed in Iran at the time of the establishment of Dar-ol-foonon one hundred fifty years ago, the peculiarities of which many Iranians perceive today to have augmented our culture, are known as the modern sciences.

Science or knowledge is a collection of the data, facts, rules, and laws of the day, part of which were transmitted from abroad to Iran after the establishment of the University of Tehran. Based on this perception, a crosssection of accounts of the scientific process reflected in scientific texts at the time is considered as science itself. This concept is still more or less prevalent in Iran, and is also based on books (texts).

Science in its modern conceptualization is a process, the outcome of which is scientific knowledge. Participation in this process is a complex task, and our familiarity with it is of recent historical origin, and started more or less at the beginning of the Iran-Iraq war. Accordingly, this concept is alien to many of our people, and our universities and research institutes are not yet administered based on this perception of science. This concept of science is not based on texts, but it creates texts.

The most prevalent concept of science in Iran is a combination of concept 1 above and some of the components of old or new knowledge. The dependence of concepts 1 and 3 on texts, and the mastery and control of the educational system in our society by "science as religious studies" (*elm*) after the establishment of the *Nezamiyeh*(s)[2] has created a robust combination from which it is difficult to escape. The word *aalem* creates the impression not of a scientist or a natural scientist, but a scholar of religions in the mind of many Iranians. Note that the Persian term for "study" is equivalent to "reading." Therefore, the literal translation of scientific studies in Persian would mean "reading science," which once more stresses the significance of texts in the prevailing concept of science in the Iranian culture. Hence, the sciences are to be read and not performed. We describe a learned person, often a man, as a person who is bookish and who has read lessons in texts. The person who takes steps on the scientific path is one who "reads lessons." The verb "study" is equivalent to the Persian verb *khandan* and implies that one should accept science as something written in books that must be "read" or "memorized."

No wonder that our precollege educational system, even the tertiary education system is based on reading and memorization. In such a system, there are no signs of creativity, production, innovation, and initiative.

This prevailing perception of science has deep roots and goes back to the tradition of the *Nezamiyeh* and old religious seminaries (*hozeh elmiyeh*). These schools used to play the role of modern universities until the new era in Islamic societies. In the twelfth or the thirteenth centuries, the teaching of natural sciences and mathematics became a part of the curricula of religious schools, and at the same time was liberated from the domains of wealthy benefactors, and institutionalized outside the courts, bringing it under the control of religious studies, which afterward were named science (*elm*). Under the prevailing historical conditions, this has led to the concepts of the utilitarian sciences (*elm-e-naafe*) versus the harmful sciences (*elm-e-zaar*) and the banning of the harmful ones, which ended in the dormancy of the latter sciences.

We should not confuse the meaning of *elm-e naafe* with the modern concept of mode 2 sciences (Gibbons et al. 2000). *Elm-e-naafe* in our tradition was the knowledge that was required by theology, subjects like a little mathematics and a bit of astronomy, which is totally different from the sciences to which we are accustomed in the modern era. The distinction between the two concepts of knowledge and science is not merely a semantic one; this differentiation has profoundly impacted the developmental policies of the country. In Iran, there is not yet any consensus on what constitutes modern science. A major part of the current debate in Iran and other Islamic countries pertaining to Islamic science and Western science is the result of misunderstanding about this differentiation. We still face difficulties in distinguishing between *elm* and science, as well as between a modern scientist and an Islamic theologian: Both are referred to by the same term in Persian, *aalem*. This difficulty, that is, the mixing of different terms and concepts, has created serious misunderstandings in our society. When an influential cleric, in discussing the relationship between science and *hozeh* in present-day Iran, considers the empirical sciences to be "defective" (Javadi Amoli 2011, 96), and at the same time writes: "if science is science, it cannot be non-Islamic" (Javadi Amoli 2011, 106), we must conclude that we are suffering from a grave misunderstanding. Accordingly, these misunderstandings and Iran's historical experiences with which we shall deal in this chapter, create a serious impediment to the development of the sciences in Iran.

A History of the Higher Educational Institutions in Iran

A knowledge of the history of higher education in Iran helps in understanding the history of science in the country. Please keep in mind that the Farsi term *daneshgah* was coined to mean university 80 years ago, even though the history of higher education in the country goes back to the pre-Islamic era. Figure 2.1 shows the evolution of the educational system from the pre-Islamic era to the present. In this figure, we see several discontinuities that have played key roles in the evolution of the sciences in the country.

Jondishahpour University, as the Persian institution of higher learning with the greatest longevity, was established by Ardeshir Papagan (180 AD-241 AD), the first Emperor of Sassanid dynasty. Even though before establishment of the Jondishahpour Uinversity Ardeshir passed away, his son and successor Shahpour opened the university in the city of Dezful in the current Khuzestan Province, seven years after his father's death. Jondishahpour operated for five hundred years up to the mid-ninth century. The next most enduring institutions of learning are the seminaries, called *houzeh* in Persian, which have functioned for the last nine hundred years and have flourished since the Islamic Revolution. These examples show how educational centers for the natural sciences and philosophy after Islam usually had short longevity, and as a result did not have opportunities to establish a persistent tradition.

Pre-Islamic Institutions

I know of no scientific institution with a tradition that goes back to the pre-Islamic era other than Jondishahpour University. This tradition is rare even throughout the entire Islamic world. The schools of Athens and Alexandria had lost their importance at the time of emergence of Islam. Its influence was mostly observed during the rule of Anushirvan, the Sassanid king. Shahpour Ben Sahl, who died toward the middle of the ninth century AD, was the last physician and head of the school before the school's tradition (knowledge-base, administrative and goverenance structure, and reputation) was transferred to Baghdad and the hospitals in the early Islamic-era. It seems this is the first discontinuity in the history of our higher education.

We do not have much information about the teaching of engineering; perhaps teaching such subjects was not customary in those days. However, we have engineering masterpieces such as Tagh Kasra Arch in Tisfoon and the water supply system complex in Shooshtar, which are still in existence.

The Islamic Era until the Formation of the *Nezamiyeh*(s)

Many books and articles on science and the history of science in the early period of Islam have been published (Soltanzadeh 1985; Kassai 1984). From these publications, we know that institutions such as *dar al elm* (house of science), *dar al kotob* (house of books), *dar al hekmat* (house

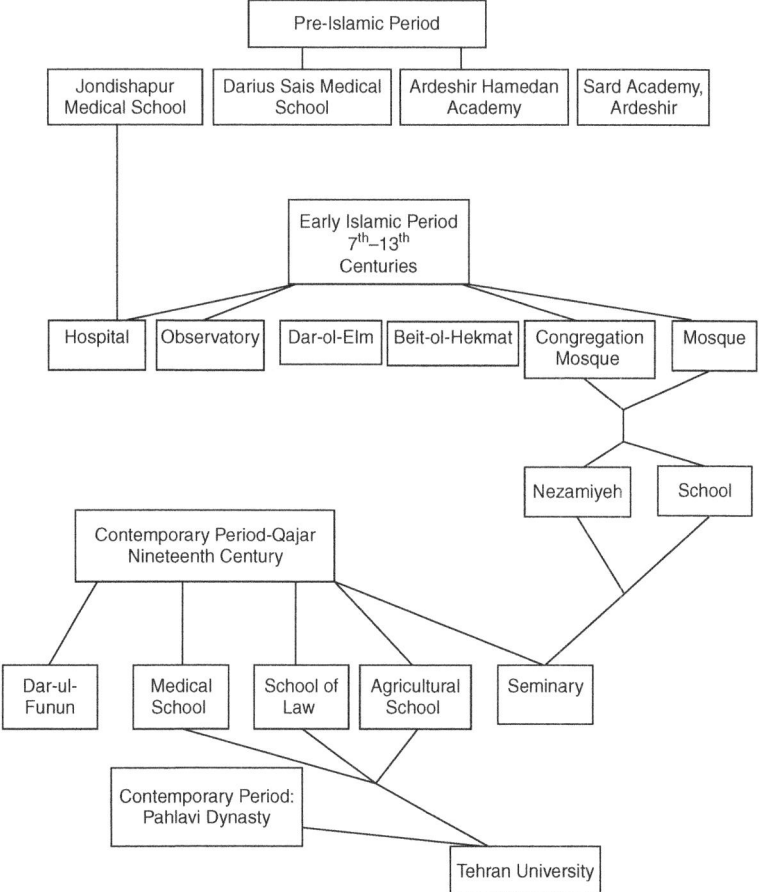

Figure 2.1 History of Science Institutes in Iran

Development of educational institutions in Iran: Jondishapur Medical School was the only pre-Islamic institution that was transformed into an Islamic educational institution. Theological seminaries are the only educational institutions still active today that are rooted in the early Islamic period.

of philosophy), *dar al hadith* (house of Hadith), *dar al Quran* (house of Koran), and Majsied Jaumah (Jaumah Mosque), and the ordinary mosques and madrassa were established beginning in the early eighth century and were active in various scientific fields. Additionally, local courts across the country, such as Samanids, Bouyids, and Ghaznavids, had libraries as well as educational institutions, and scholars who were financially dependent on them. Starting in the Seljuk dynasty and at the time of establishment of the *Nezamiyeh*(s), a new era of learning and scientific activities started in Iran. The era was marked by a decrease in the diversity of scientific activities, and limits were imposed that restricted the exploration of the natural sciences and philosophical discourse to the officially sanctioned framework of the *Nezamiyah*(s). As a result, a new discontinuity in Iranian scientific institutions emerged. Science education was reduced to theology, and the term "madrassa" was only used for religious schools. It is for this reason that from the thirteenth century onward, the discontinuity in the sciences and scientific centers began, peaked in the Saffavid and Qajar dynasties, and continued until the time of the modern transformation.

Recent Era: The Period from Naseriera to the Islamic Revolution

With the creation of modern educational institutions such as elementary and high schools, *Dar-ol-fonoon* (house of technology), and medical, legal, and agricultural colleges, and then their merging into the University of Tehran, a new era in education began in Iran (see Figure 2.1). Figure 2.1 clearly shows that the institution of the university (*daneshgah*) is not linked to any traditional institutions of learning in the country. One exception is that the method of teaching of the sciences was influenced heavily by theology, such as the traditional science education after the *Nezamiyeh*(s) at *hozeh elmiyah*, or theological seminaries. Attention to the discontinuities is important in the analysis of the current and future states of the sciences in Iran.

Analysis of the Evolution of the Sciences in Different Eras in Iran

In my book *Iran 1427*, based on the social dynamics perspective, I analyze the state of the sciences in Iran during different historical epochs after the advent of Islam (Mansouri 2007c). One may accept the analysis or ignore it, because it belongs to an amateur historian of the sciences! However, what is important is that taking into account the social dynamics and knowledge of the scientific evolution in the world, a subject that our historians have been inattentive toward thus far, I have distinguished among various eras, which had fundamental differences and impacts on modern history by viewing the eras from perspective of the concept of modernity and of knowledge-based societies. Paying attention to these differences enables us to better understand the future and the dynamics of scientific development, and allows for a better understanding of the development of the sciences in Iran in the past. Here I would like to extend this categorization to pre-Islamic Iran as well as to the Islamic "Golden Age," until the establishment of the *Nezamiyeh*(s).

Science in the Pre-Islamic Period

Our main information about the state of the sciences starting in the pre-Islamic era goes back to the Sassanid dynasty (AD 224–AD 651), although the information on medical sciences, including the transfer of methods of surgeries from Persia to Greece and the reopening of Sais Medical College by Darius, goes back to the Achaemenid Empire (BC 550–BC 330) (Soltanzadeh 1985). However, the various schools in the Sassanid Empire have given this era a particular historical significance that has emerged from the safeguarding of traditional education for more than two hundred years after the emergence of Islam. The most famous institution of this era is Jondishahpour University, which is considered as the symbol of the development of the sciences in Ancient Persia. The fact remains that at the time of the emergence of Islam, the Alexandria School was in decay, and in the Eastern Roman Empire, schools were less prevalent. All of these circumstances occurred at a time when Jondishahpour University was succeeding in attracting medical scientists, mathematicians, and astronomers from the East and the West to the university.

The activities of Jondishahpour University for five hundred years were a sign of the development of the sciences in Sassanid Persia. However, we should not ignore the fact that the official government religion during this period had complete dominance over all cultural activities, including the schools, in the empire, which imposed limits on rational thought. Perhaps the lack of adequate growth in philosophy during this epoch was due to the limits imposed on rationalism by the official religion, which limited critical thinking and philosophy to mythology and religious ideas. It is clear that this restriction did not impede the progress of the medical sciences, engineering, or astronomy and astrology.

Both the water system complex of Shushtar, which is considered one of the major engineering achievements of the time and which has lasted to the present day, and the Arch of Kassra (Taugh Kassara) in Tisfoon refute the limitations imposed by religious thought: apparently the dominance of religion and the constraints on speculative thought did not contradict medical and engineering practices. However, these limits had profound cultural impacts on the society, on rationalism, and on the prevailing definition of the sciences in Iran. For this reason, one must be careful in claiming the Sassanid period as an era of scientific progress in ancient Persia, as some historian of the sciences are inclined to do.

Much evidence points to the subordination of rationalism by jurisprudence and theology during the Sassanid Empire. The schools at the time were managed by *mogh*(s) (Zoroastrian priests). Accordingly, in discussing science in the period before Islam, one should pay attention to the deep impact that religion had on science, and keep in mind that science was considered a part of religious thought and was bound by it.

The Golden Age 1: The Era of Infrastructure: From the Seventh to the Tenth Centuries

The first century of Islamic rule in the Sassanid Empire corresponds to notable cultural and social transformations, when no one would have expected important scientific events to occur. What we could say a posteriori is that in spite of the continuing activities in educational centers such as Jondishahpour, given the domination of the Persian culture by the Islamic developments, conditions did not exist in the country for new scientific institutions to be established. Under such conditions, even the continuation of the activities of the existing centers could not be assured. The important point is that language, on which thinking as such is based, did not have the ability to create the conditions for the growth of thinking in the newly born Islamic society. However, in the southern Arabian Peninsula, a civilization was formed in the first millennium BC, and the southern Arabic languages called Moayani and Sabaie (*Homayri*) (Dabir-Moghaddam 2011), and 29 written alphabets had appeared and were in widespread use. In the north of the Peninsula where Islam emerged, reading and writing were not prevalent, and only a handful of Arabs were literate. Under such conditions, it is meaningless to speak of scientific transformation. Of course, without the evolution of language, the evolution of thought and the sciences is unimaginable. From this perspective, what Sibawayh did with Al-Ketab is without doubt a turning point, not only in science in the Islamic world but also in the history of scientific evolution and linguistic history of mankind.

Abu Bashar Farsi Bayzavi, known as Sibawayh, was born in Bayza, the Fars province of Iran, in 763 AD, and he died in 795. He is unanimously considered by linguists as one of the founders of a linguistic tradition and one of its best representatives (Dabir Moghaddam 2010 and 2011). One of the well-known linguists before Sibawayh was his teacher, another Persian with the name of Khalil Ben Ahmad Farahidi (717–785). Farahidi was a descendant of Iranians who were sent to Yemen during the reign of Anushiravan, the Persian king. That scholars engaged in language studies and created a language capable of expressing scientific concepts in this early period may be inferred from both the number of linguists of the early Islamic period and from the number of workshops pertaining to discussions of linguistic problems that were held (Dabir-Moghadam 2011). It is interesting to note that in 960 AD, at least 265 linguists and philologists were cited in oral history, the most famous of whom were Persians (Itkonen 1991; Fakhri 1970, 49–50). Attempts by Sibawayh to prepare an Arabic language capable of expressing scientific concepts, first through translations and then through the production of scientific output, are contemporary to the Abbasids rise to power, a dynasty that came to power from the Eastern part of Islam, Iran.

There are no signs of efforts to institutionalize the sciences in this early Islamic period during the Omavadis time, and only reports of the isolated teaching of children and youths are available. Freedom of thought, reasoning, and progress in the sciences—at the time when language did not have adequate growth—were not problems of interest to the rulers, although the start of freethinking and discourse on different doctrines like *qadarites* (free will) and *jabries* (determinism) began in this period.

With the Abbasid dynasty in the middle of eighth century, which adopted the Persian culture of the Eastern Islamic territories, the institutionalization of the sciences in the Islamic world began. Mansur Abbasi, who was the founder of the Bani Abbas dynasty, established his capital in Hashemieyh, a city located near Koufeh. However, after encountering revolts, he moved his capital to a Persian village named Bagh-Dad in 758 AD, and established a city with the name of Dar-Al Salam, which later became famous as Baghdad.

It is interesting to note that the Abbasids, in addition to being educated in the Persian region of Khorasan, established their capital city, Baghdad, near the previous Sassanid capital, Tisfoon, an area where the Persian culture was dominant. The area was called "Irak" in the Old Persian language, meaning "low country" (Rashed-Mohassel 2011). With the growth of the Arabic language and its rationalization by two Iranians, first by Farahidi and then by Sibawayh, the Arabic language gradually spread in this region, which was populated mostly by Iranians, and from there it penetrated into other Iranian regions. The date of establishing the city of Baghdad was determined by an Iranian astronomer named Nobakht, who worked in Mansur's court. The year 770, in which the translation of an Indian book on mathematics and astronomy was begun, could be called the start of the "Golden Age" of Islamic civilization. The first scientific institution in the time of Maamun (in 884), with the name of Bayt Al-Hekamat, was established in Baghdad.

It is clear that linguistic activities in the eighth century were extensive and created suitable conditions for the translation of books from other languages into Arabic, as Arabic became the language of the sciences in Iran as in other Islamic countries, although Iranian scholars wrote in Farsi too.

This era is, at the same time, marked by the rise of different local dynasties within Iran, before the Saljuk dynasty appeared in the second half of the eleventh century. Large public libraries, called Dar-Al-Kotob and Khazanat-ol-Hakmeh, were established, followed gradually by Dar-Al-Ulum, or houses of the sciences, toward the end of the eighth century. After a while, local dynasties supported and established similar institutions. This century may be considered as the century of the establishment of scientific institutions such as libraries, schools, and Dar-Al-Ulum. Of course, as we have seen, during this period, Jondishahpour University and its traditions were moved to Baghdad, and a new era in institutionalizing the natural sciences was established.

The Golden Age 2: The Age of Enlightenment, the Tenth and Eleventh Centuries

Starting in the tenth century, along with the institutionalizing of the sciences, the ruling dynasties in Iran found more freedom and engaged in competition in establishing scientific centers and attracting scientists. Freedom of thought found more opportunities also. The Abassid dynasty initially established freedom of discourse and thought; however, beginning with Motavakel (845–860) and the prohibition on discourse and debates, freedom of thought disappeared. These conditions led to the migration of many freethinkers to the regions that were ruled by Iranians, resulting in the growth of the sciences in the courts of freethinking Iranian rulers. The policy to support wisdom and intellect peaked in Buyid's (930–1018) court to the extent that not only engineering and medical science practices grew, but freethinking per se, such as natural philosophy, grew also. No wonder this era is known as the era of Islamic Enlightenment (Kremer 1992). The unmatched role of the Iranian rulers and their prejudice-free courts generated opportunities for freethinking and for institutionalizing the sciences, and after the establishment of the Arabic language as the language of science, and with the existing scientific tradition in the region of the Sassanid Empire, a rare situation emerged in Islamic civilization that was never repeated again. Whenever we look at the achievements of the Muslim civilization, we come across names such as Farabi (880–949), Avecina (980–1037), Abu-Rayhan Birooni (973–1048), and Khayam (d. 1123): all of these giants of the human sciences lived in a two-hundred-year period between the ninth and the twelfth centuries. Even though Sibawayh and Kharazmi (850–780) lived earlier than this period, one must consider them a part of the Golden Age.

There were many influential scholars during this time. For example, there were at least 15 high-ranking astronomers who worked with Abdulrahmaun Soofi at Buyids Observatory near Shiraz in the second half of the tenth century. We should keep in mind that this number is much higher than the number of astronomers who are working in present-day Iran 1100 years later!

It is informative to name a few mathematicians and astronomers of this period who were the founders of mathematics and astronomy at this time human history. Among them were Abu Hanifeh Deanvari (850–929), Abol-Vafa Bouzjani (938–997), Koushyar Gilani (941–1009), and Abu Mahmud Fajandi (no records of birth or death dates, but he is believed to have lived in the tenth century). According to Abu Rayhan Birooni, Fajandi was an astronomer with a rare talent for building astronomical instruments, and the Fakhri sextant that was used in an observatory at the top of Tabark Mountain near the city of Ray was his construction (Abdulla-Zadeh 2008).

Beginning with the rule of the Ghaznavi Turks, the local rulers, as the supporters of free thinking and the creators of the centers where persons of reason and science gathered, lost their ability to do so, and dogmatism and the persecution of dissidents began in parts of Iran. When Toghrol Saljuk entered Baghdad in 1053, and overthrew the Buyids dynasty, which ruled in this city, the era of Islamic Enlightenment ended. This is the start of the decline of Islamic civilization, even though this is the same time when the Saljuk Empire covered wide territories.

The Era of the Decline: The Age of Transition to Decline from the Twelfth to the Fourteenth centuries

The conquest of Baghdad by Toghrol in 1054 is considered as the starting point of this era, even though the Ghaznavi Turks had begun this process earlier in the eastern part of Iran. Beginning toward the end of the eleventh century, along with the empowerment of the Saljuks, who had powerful yet narrow-minded prime ministers such as Abu Nassar Mansur Ibn Muhammad Kondori and Khajeh Nezam-ol-Mulk, freedom of thought and religion was eliminated. Khajeh Nezam-ol-Mulk, who was a follower of Shafei, which had an Asharite doctrine, and who considered this branch of Islam to be pure and correct, placed all *Nezamiyeh*(s) and all other schools in the empire they ruled, from China to the Mediterranean Sea, under the control of the scholars and followers of this "pure faith." Accordingly, people in general and the scholars in particular during these times lost their ability and power to engage in critical thinking, which set the stage for the decline of the sciences and rationality in Iran and all other Islamic countries.

Nezamiyeh(s) were the symbols of the confinement of freethinking. Perhaps to combat the Ismailiis, who were considered a terrorist group whose cultural activities directed against freethinking, Khajeh Nezaum-ol-Mulk created schools in many cities in the empire. In practice, whenever he found a person of "scientific fame" (a theologian), he created a foundation for the establishment of a school under that person's leadership. He was never interested in persons of fame in the natural and medical sciences, or even in any Islamic theology in general, but just those of Shafei or Hannafi.

In these schools, the teaching of philosophy and the rational sciences were banned, and teachings were confined to religious studies only. Intellectual discourse and religious debates that were free prior to this era often led to the destruction of the schools and even the death of some debaters. For example, in 1075 when Imam Abu Nasser Ghosheiri debated in the Baghdad *Nezamiyeh* in support of the Asharites' ideas and against the ideas of the Hanbalites, a mob of Hanbalites attacked and killed a number of the Asharites' followers. In 1094, in Nayshabour, a clash between the followers of the Shafeits and the Hanifis on one hand and the followers of the Kerameyah on the other also resulted in the destruction of Kerameyah schools. Moreover, in 1159 a clash between the Hanafis and the Shafeites resulted in the destruction of 25 schools and 5 libraries in the city of Nayshabour as well.

In such ways the empowerment of the Saljuks' religious intolerance prevailed, and management of the affairs of the country was combined with militarism and extensive control of local rulers through security mechanisms. Instead of promoting freethinking and rational discourse, strict limitations and the creation of impediments for learning, even of the natural sciences, prevailed. Perhaps the historical regression of the Golden Islamic civilization was an unintended, destructive consequence of this type of political governance of the country, from which the country could not recover for almost one thousand years.

This was the damage that the Saljuk dynasty, with a centralized, prejudiced governance system, inflicted on Iran and all Islamic societies. From that time on, with few exceptions, Iran never witnessed the emergence of influential scientists who advanced the sciences in the world. We can name some isolated breakthroughs in the natural sciences in Iran and other Islamic countries as the exception. (For details, see Sabra 1987).

The Era of Decline II: The Age of Ignorance and Consternation, 1400–1920

The age of ignorance coincides with absolute stagnation and decay in the Islamic societies. Khajeh Nezam-ol-Mulk, who was a contemporary of the great antirational religious scholar Mohammad Ghazali, is the greatest political thinker of the time of the Turkmen rule in Iran. In his political thought, he paid attention only to the acquisition and maintenance of power. Instead of confronting the growing influence of the Christian West in the Islamic territories, he was preoccupied with controlling the local opponents, to whom he referred as "dogs." Khajeh Nezam al Mulk's book, *Syasat Nameh* (Policy Papers) (Khajeh 1985), was written in the early period of antirationalism. It seems that two hundred years of scientific decline culminated in a new age of "ignorance," which coincided with the first stage of the Iranian government's interactions with the West and with absolute intellectual stagnation and antirationalism. Even political philosophy was reduced to political ideas based on theology. Moreover, the remnants of Khajeh Nezam-ol-Mulk's era of rationalism were replaced with a special interpretation of religion.

In the late fifteenth century, Iranian scholars were deeply engaged in how to rule based on religion, and no signs of historical analysis in their political thought were present: religious symbolism dominated their thinking. The dominance of religious symbolism in this era, which had its roots in the purging of the natural sciences and rational thought from the *Nezamiyah*(s), paved the way for this era of ignorance. During the Safavid rule, clergymen Mohsen Faiz Kashani and Muhammad Bagher Majlesi warned Muslims that they should avoid knowledge based on reason and rationality. Mirza Qumi, the head of the Qom seminaries in the early twentieth century, who considered the famous philosopher

Mullah Sadra's blasphemy transparent, did not allow the students at the seminary to study any subject other than Islamic jurisprudence. It is hard to imagine that in such an era scientific progress could occur. Accordingly, the era after Muhammad Ghazali and the *Nezamiyah*(s) must be considered as the period when the definitive decline of the sciences in Iran started. Toward the end of this era, which may be called the Age of Consternation, especially after three periods of modernization by Abass Mirza, Amir Kabir, and Sepah Salar, the bewilderment and inaction of Iranian scholars gave way to thinking about alternatives to progress, but at the same time to extremely simplistic thinking: some considered the philosophers of the Qajar time as the sun, in contrast to Western philosophers such as René Descartes, whom they saw as Mira, a small star in the constellation Ursa Major. This exaggeration is a reflection of simplistic thought, and is another sign of the ignorance in this period (Mansouri 2007c).

In sum, it must be stated that the light of science was turned off during this era. The darkness was accompanied by false pride and simplistic thinking, which resulted from the culturally perceived, imaginary presence of scientists in the Golden Age of Islam, whose influence on science and policy planning in current Iran must be considered

At the end of this era, particularly starting at the time of the opening of Dar-ol-foonon, murmurs about modern scientific institutions and talk about the Persian language as the national language of science were prevalent, a subject that I have dealt with in my book *Terminology in Iran and the World* (Mansouri 2011).

The Age of Transition to Awareness or Age of Self-Realization

This era still distanced itself from awareness of the concepts of socioeconomic development, scientific development, and their historical dimensions, even though one could differentiate among periods, such as the years of Reza Pahlavi's rule (1925–1941), and the periods 1942–1953, 1953–1965, and 1966–1979. The graduates of this era, who had deeper insights than those who were educated in the West during the Age of Consternation, still had a crude perception of the modern sciences and technology. No exquisite, influential articles in the scientific and technical fields were published in these times. The management of the sciences at all levels was very rudimentary. The ethics and behaviors of the Age of Decline were openly prevalent in the scientific centers. Apparently, the long periods of neglect and then bewilderment had created deep fears in the scholars of the previous generations, so that engaging in concrete, profound, and deep modern thought seemed impossible for them. Or else, based on a superficial familiarity with Western expertise, they became so excited that they never pressed for a deeper understanding or continuation of the education in Western expertise that they had acquired. This created an unpleasant phenomenon in the country, which I have called "un-grown-up" personalities (Mansouri 2007a).

Our very talented young students, especially those who could not attend reputable universities abroad, and who were educated in Iran, were never able to engage in in-depth research in any field. On the other hand, these talented young people gradually were transformed into the elders who had not grown up. This phenomenon, which is the pathology of our society, by definition could only affect the best talents of the country. As a result, one generation of our most talented youth were despoiled, a generation that could have potentially engaged in serious, meaningful research. Moreover, a kind of personal false pride, which had not been disconnected from the false pride of the era of ignorance, emerged among those who had returned from the West, individuals who failed to consider their knowledge and science as incomplete and who were not able to carry the burden of educating the future generations. They continued the old tradition that the "teacher knows all" and contributed to the helplessness of the "un-grown-up" talents.

Starting in the early 1960s, particularly in the 1970s, awareness of the uneducated behaviors of academics appeared. Perhaps one of the reasons for the establishment of Shiraz University and the Sharif University of Technology, or the establishment of research councils in Iran can be found in the collegial anti-intellectual conduct of the era. Certainly, as a result, the revolt against collegial antiscientific conduct after the Islamic revolution flourished.

The Age of Awareness: 1980–Present

After the triumph of the Islamic Revolution, the energy of a large part of the society directed toward the construction of the education system was released into an environment in which everything seemed possible. Iranian nationalism and the desire for national and cultural independence from the West, a dependence that had lasted for the better part of the Age of Consternation and the transition to awareness, almost two hundred years, were the motivations for this release of energy. The intellectual environment immediately after the revolution was very open, and created the fertile, nourishing grounds for development of innovative ideas. Numerous movements to reduce the impact of the Ages of Ignorance and Self-Realization began, and a great deal of knowledge about the strengths and weaknesses of the country was acquired. If Pahlavi claimed that Iran would reach the "entrance to civilization" in five years, and the Iranian technocrats at the time believed him, or pretended that they believed him, in postrevolutionary Iran, few people doubted the hollowness of such a claim. The frank acknowledgment of backwardness in all affairs, and at the same time the strength of the Iranians, forced academics, researchers, and seminary students to find solutions to remedy the country's problems. Naturally, many of these solutions were of the trial-and-error kind; nevertheless, they were effective and led to noticeable progress in many cultural and scientific fields, even though Iran still requires much time to grasp the modern perceptions of sciences and modernity and to develop domestic capabilities in these areas. The beginning of the era of awareness does not mean, however, that the society as a whole has become aware.

We may cite much evidence for the start of awareness in Iran; some is qualitative and some is quantitative. Some of the qualitative evidence and the quantitative evidence are cited in Mansouri (2007c). Table 2.1 shows some of the evidence. The absolute value of the figures for 2011 show that Iran has closed the threshold gap for the definitive development of the sciences. This threshold is defined as 25 percent of the norm for the industrialized countries.

In addition to the quantitative indicators, we must pay attention to the growing discourse indexes in all aspects of science, technology, and education that were never seen before in modern-day Iran. The awareness of science and technology policy making, including the extensive activities of the National Research Council of Iran after the revolution,

Table 2.1 Indicators of Educational-Scientific Development, 1979–2012

Indicator	Number in 1979	Number in 2012	Absolute Percentage increase	Percentage increase relative to population
Universities	15	120	8 fold	4 fold
College enrollment	150,000.00	4.5 Million	30 fold	15 fold
Articles published in International journals	400	20,000.00	50 fold	25 fold
Number of books published	150	70,000.00	500 fold	250 fold

has become as extensive as debates about "Islamic studies" and the Islamization of the universities. Various forces and groups are involved in this discourse and its consequent awareness. During the first two decades after the revolution, academics, manufacturing, the service sectors, and the defense industry were involved in the debate, with the result that they became an effective force in propelling science and technology.

If we divide the cultural sphere into *Hozavi* (seminaries) and general (modern studies), we may obtain a clearer picture of how these two sectors have influenced the sciences and how they have been influenced by the sciences in the country. The general educational sector, being enthusiastic about learning modern science and technology, has been both a force for the promotion of science and technology, and at the same time has been deeply influenced by it. However, the traditional education system of *hozeh*, has looked with deep suspicion on science and technology, and particularly on the modern sciences. During the first two decades after the revolution, only this sector of higher education in Iran has been influenced by science and technology; however, it has been silent on that front. That is, it has been influenced by science and technology, but has not been active in the discourse. The influence of science and technology on the seminaries has been mostly due to the translation of books or the teaching of instructors who have been trained in the modern social sciences and humanities. Moreover, since the seminaries rarely were in contact with first-rate scholars or first-rate writings, and since they were not exposed to the modern natural sciences and the cognitive foundation of these sciences, on which the modern social sciences are also based, a misinterpretation of the modern sciences has developed in this sector. For this reason, the influence of the modern sciences on seminaries is rather distorted.

New technology and its influence on the seminaries must be examined in a different way. Subsidies, all sundry educational supports, and the higher education industry with all its derivatives have penetrated the seminaries, and from this perspective, this phenomenon has given the seminaries an appearance of modernity. The seminaries are recipients of the latest information technology and the latest methods of testing, and the books for entrance-exam tests for the academic programs at the seminaries are widely available. It is too early to assess the impact of the new technologies on the transformation of seminaries, but its influence is certain. Perhaps contrary to expectations, software and hardware information technology—not the modern social sciences—will bring about a major transformation in the seminaries and then prepare them for a deep encounter with the cognitive foundations of the modern social sciences.

However, it appears that the seminaries have found opportunities to break their silence in confronting the modern sciences. Certain members of the traditional religious schools, along with a number of politicians, have expressed their serious objections to the natural sciences as well as the humanities. As examples, we cite two statements from the theologians in *hozeh*. First: "What is taught at the universities are carcasses that are incapable of becoming Islamic" (Javadi Amoli 2009). Second: "In such conditions... we need to connect the sublime, Islamic thought with fundamental scientific subjects such as physics, chemistry, and mathematics in ways that the students in these fields search for the Unity of God and other tenets of Islamic teaching that are the most fundamental scientific principles, in the scientific subjects, and by solving problems and using formulas learn ethical, Islamic lessons" (Nabavian 2009).

The opposition has gained momentum recently, and uses all political instruments in this pursuit. I have discussed the details in my book (Mansouri 2012; It appears that the confrontation between *hozeh* and universities in recent years will determine the relationship between science and religious studies in the final analysis. The religious domination of wisdom, and not their synchronization, will be made clear after two thousand years in Iran. For this reason, one should not fear the recent developments, and should consider them a blessing.

The impact of manufacturing and the service industries on the development of the sciences in Iran should be kept in mind. These sectors are isolated from the current controversy between the modern concept of science and the traditional notion of science as it emerges from the seminaries. However, due to the acute needs of the industrial sectors of the economy, particularly the defense industry's dependence upon the sciences, it seems that over the long run, these industries will determine the outcome of the battle between the universities and *hozeh*.

Conclusions

Before the Islamic Revolution, after two centuries of acquaintance with institutions of modern civilization, including scientific, economic, and political institutions, and after a half-century of the existence of universities, Iran did not have a full comprehension of the roots of the modern sciences, or understand the concept of modern sciences, nor had it gained insight into the relationship between human beings and nature. For many Iranians, there exists a psychological obstacle, which makes acceptance of the reality too difficult. This psychosis is due to the greatness of the Sassanid Empire and the Golden Age of early Islam, which inhibit them from awakening from their eight-hundred-year slumber, which started after the establishment of *Nezamiyeh*(s) in the Saljuk dynasty, so that they can realize they have been separated from the caravan of the evolution of human thought. For this reason, many Iranians have difficulty in understanding the meanings of modern science and technology. This problem has reflected itself in the process of science and technology policy making, which has an important role to play in modern times.

Even though, toward the last decades of the Shah's rule, an institute with the name of the National Council of Research was established in Iran, it is difficult to consider that the activities of the Council were based on science policy. Science policy making in Iran started after the establishment of the Islamic Republic. Furthermore, the psychological block in relating to the modern conceptions of science and technology is not unique to Iran: many other Islamic countries face the same impediment (Natali 2011, 74).

The current confrontation between modern science and traditional theological studies that has heated up in Iran recently will determine the outcome of a two-thousand-year struggle between scientific rationalism and religious rationalism. Whatever the outcome, it would be closer to the conditions at the time of the Buyids and the Golden Age of Islam than the other eras of our civilization.

Notes

1. The city of Najaf in Iraq was and still is a prominent center of Shiite Islamic studies.
2. *Nezameyih*(s) were old schools in Persia that were attended by seekers of knowledge. These schools, like many modern-day universities, provided living quarters, and the attendees were financially supported, usually by the tutors of professors, who were high-ranking religious leaders and scholars with their own following or *omeh*.

References

Dabir-Moghaddam, Mohammad. 2010. *Dastour, Vizeh-Nahmeh ye Farhangestan* "Moarafi va naghade ketabe makateb zaban shenasi novin dar gharb" [Introduction and critique of the book on linguistic schools in the west] No. 5, pages 3–43.

———. 2011. *Zaban-Shenaasi Elmi Sharghi* [Linquistics as an eastern science].
In *Jashn-Nahmeh ye Salim Neysari* [Honoring Master Salim Naysari] (in Farsi), edited by Hadad Audel, Gholamali, 165. Tehran: Farhangestan e Zaban va Adab e Farsi, Nashr e Aasaar.
Day, Natalie, and Amran bin Mohammad. 2011. *Malaysia: The Atlas of Islamic World Science and Innovation*, Country Case Study, No. 1, San Francisco: Creative Commons. Accessed August 22, 2012. http://creativecommons.org/licenses/by-nc-nd/3.0.
Fakhry, M. 1970. *A History of Islamic Philosophy*. New York: Columbia University Press.
Gibbons, Michel et al. 2000. *The New Production of Knowledge*. London: Sage Publication.
Hendou-Shah, Sanjar Nakhjavani. 1965. *Tajareb-o-Salaf* [predecessor's experiences] (in Persian). Tehran: Tahouri Publishing.
Homaii, Jalalelddin. 1977. *Ghasali Naahmeh* [Ghazali's Letter] (in Persian). Tehran: Foroughi Publishing.
Hoodbhoy, P. 1992. *Islam and Science: Religious Orthodoxy and Battle for Rationality*. London, U.K: Zed Books.
Itkonen, Esa. 1991. *Universal History of Linguistics*. Amsterdam: John Benjamins Publishing.
Javadi, Amoli. 2011. *Gozideh ii az Manzelat e Aghli dar Hendeseh ye Marefat e Dini* [A selction of degree of rationality in geometry of religious knowledge]. Daftar e Nashr e Qom: Ma'aaref Publication.
Kasaii, Nourollah. 1984. *Nezamiyeh Schools and Their Social Impacts* (in Persian). Tehran: Amir Kabir Publishing.
Khajeh, Nezamolmolk. 1985. *Seyaer-ol-Molouk* [behaviors of kings] (in Farsi). Tehran: Elmi Farhangi Publishing.
Khorshid-Feyzollah, Abdollah-zadeh K. (2009). *Koushyar Gilani*. Translated from Russian to Persian by Parviz Shariyari. Tehran: Entesharat e Elmi Farhangi Publishing.
Kraemer, Joel L. 1992. *Humanism in the Renaissance of Islam: The Cultural Revival during the Buyid Age*. New York: E. J. Brill.
Mansouri, Reza. 2012. *Terminology in Iran and the World* (in Farsi). Tehran: Farhangestan e Zaban va Adab e Farsi.
———. 2007a. "How an Obsolete Concept of Science Impedes the Development of Islamic Countries." *Forum on International Physics Newsletter*. America Physical Society, December. Accessed August 22, 2012. http://www.aps.org/units/fip/newsletters/200712/iran.cfm.
———. 2007b. *What Should I Do with Iran?* (in Persian). Tehran: Kavir Publishing.
———. 2007c. *Iran 1427* (in Farsi). Tehran: Tarh-e Nou Publishing.
———. 2011. *Iran 21/15* (in Farsi). Tehran: Ettelaa'aat Publishing.
Nabavian, M. 2009. "Unprecedented Statements about Teaching Islamic Doctrines at the Universities." (in Farsi). May 7, 2009. Accessed August 22, 2012. hhttp://www.fardanews.com/fa/news/82237.

Sabra, A. I. 1987. "The Appropriation and Subsequent Naturalization of Greek Science in Medieval Islam." *History of Science* 25: 223–243.
Sardar, Z. 2007. "Beyond the Troubled Relationship." *Nature* 448: 131–133.
Serageldin, I. 2007. "Islam and Democracy." Address to a seminar organized by the Norwegian Ministry for Foreign Affairs and Religions for Peace Conference. Oslo, Norway, 20 June 2007
Shi, Y. 2001. *The Economics of Scientific Knowledge: A Rational Choice Institutionalist Theory of Science.* Cheltenham, UK: Edward Elgar.
Soltanzadeh, Hossein. 1985. *A History of Iran's Educational Institutions from Ancient Times up to Dar-ol-Fonoon* (in Persian). Tehran: Agah Publishing.

CHAPTER 3

From Developing a Higher Education System to Moving toward a Knowledge-Based Economy: A Short History of Three Decades of STI Policy in Iran

Mehdi Goodarzi and Soroush Ghazinoori

Introduction

A cursory look at the history of science and technology development in Iran shows that policy making for scientific development, research, technology, and innovation appears to be a modern phenomenon in the society. Such a review indicates that a robust plan for development of the science and technological system, in the form in which it exists in the West, did not evolve in Iran until about one hundred years ago. However, starting with establishment of diplomatic relationships between Iran and the West during the Qajar dynasty (1785–1925), cultural exchanges between Iran and the West began. As a result, many Western models of education and cultural values found their way to Iran, and Iranian government officials gradually adopted these Western models. Adoption of the Western models gradually lead to the modern-day educational-policy planning system and eventually played an important role in the evolution of the national system of innovation in Iran.

Since educational planning in Iran, as in other countries, is considered the main foundation for the development of science, technology, and innovation (STI). Furthermore, given that planning for the educational system takes place within an economic, social, and cultural framework, we will discuss a short history of the evolution of educational planning and science policy making in the country in this chapter. Furthermore, we will examine the research and technology policymaking in Iran as well as review the most important documents that have been produced relating to these policies. Finally, the challenges and prospects of advancing policy making for science and technology in Iran will be examined.

A Short History of the Development of Science, Technology, and Innovation in Iran

Formation and Development of the Education System in Iran before the 1979 Revolution

The history of formal university education in Iran goes back to 1,700 years ago when Ardeshir Papagan (180–241 AD), the first Emperor of the Sassanid dynasty, commenced the formation of Jondishapour University.[1]

However, a closer look at the history indicates that until 1848, there was no sign of an important, noticeable transformation in the traditional educational system. As usual, the parents with adequate financial means would send their children to a *Maktab* for their schooling. A *Maktab* was an elementary school where a teacher who was educated in Islamic studies taught the children. The male teachers were called *Akhoond* or *Mirzah*, and the female teachers were called *Molabaji*.

It should be noted that throughout much of Iranian history before the Pahlavi dynasty, education in Iran took place within *Maktabs* with teachers who were trained in Islamic seminaries (*Hozeh Elmiyah*). However, with the opening of Dar Al-fonoon, the first European-style college in Iran by Amir Kabir, the famed, reformist Iranian prime minister (1848–1851) during the Qajar dynasty, and the hiring of foreign instructors to teach technical subjects at the college, the country's educational system went through a fundamental transformation.

We should note that the goal of establishing Dar Al-fonoon was to introduce European science and modern technology to the Iranian people of that time as the first step in the development of science and technology in the country. This college was a part of the royal court. The head of the college reported directly to the shah and had full autonomy in managing the college's affairs; however, the instructors at Dar Al-fonoon were hired and fired with the shah's permission. The shah decided who was admitted to the school, and the state provided scholarships to those who gained admission. Upon completion of their studies, the students were employed by the government (Institute for Research and Planning in Higher Education 1976).

The Ministry of Sciences was established for the first time in Iran in 1855 for the purpose of developing education and managing secondary schools. Article 19 of the addendum to the Iranian constitution, which was passed in 1906, made universal elementary education mandatory for all children, required that all public and private schools be established according to the requirements of the Ministry of Sciences, and placed all schools under the administrative control of the Ministry (Ferasatkhah 2000).

On the other hand, the establishment of the University of Tehran, the first university in modern Iran, in 1934, is one of the most important developments in the Iranian education. This period may be considered the beginning of a higher educational system in Iran.

After the University of Tehran was set up and its faculties were expanded, a number of institutions of higher education were established in the major provinces in the country. It should be noted that following the establishment of the University of Tehran, the University of Tabriz was established in 1947, and by passage of the Law of Establishment of Universities, a positive trend in the creation of other universities began. A number of other universities, with emphasis on medical and agricultural colleges, began in 1949. By the time the Islamic Republic came into existence in 1979, there were 245 institutions of higher education and 86 research institutes in Iran (Entezari 2009).

In summarizing the educational policies of the government during this period, we note that the first wave of Western-type modernism reached Iran during the Qajar rule (1794–1925 AD), when the despotic rulers of this dynasty became fond of Western science and technology. Exposure to the Western sciences caused the Persian policy makers to adopt reforms in the country's educational system. Moreover, starting in the early nineteenth century, for a period of a hundred years, the development of modern education as an essential element of science policy became part of overall socio-economic development efforts in Iran. It appears that whenever the development of the country emerged as an issue in the minds of government officials in these times, political and economic conditions permitting, then science policy was uniformly adopted. Sending Iranian students to Europe for education, opening Dar-Al-fonoon College, and establishing universities, as discussed above, are manifestations of these policies.

From the 1950s until 1977, the utilization of foreign technology as a part of the medium-term policies for technoeconomic development, and the government's purchases of advanced military hardware from the West, especially from the United States, were the top priority of policy makers. At the same time, a lack of coordinated research and development (R&D) activities, and the failure of the government to adopt a special technological policy that would concentrate on the technological development of the country were prominent. Furthermore, we would not be misrepresenting the reality to say that outside of plans for industrial development, comprehensive policies for the absorption, localization, and dissemination of technology did not exist, and that the country simply transferred technologies in the form of turnkey factories during this period. For example, the steel mill in Esfahan, which was constructed by the former Soviet Union, or the Iran Khodro car assembly lines, which were licensed from a British automotive company, Talbot, are examples of such turnkey technology transfers.

The First Decade after the Triumph of the Revolution (1980s): Reconstructing the Higher Education Infrastructure

With the triumph of the Islamic Revolution in 1979 and establishment of the interim government, the chaotic state of the Iranian universities and higher education institutions—after one and a half years of struggles by the students and instructional staff against the shah's rule—were ostensibly returned to normality. However, the political activities of the students continued in a different fashion, and the universities were transformed into recruiting grounds for all sundry political parties, which made it all but impossible for students to learn and acquire knowledge.

In March 1981, in light of these developments and the ongoing cultural transformation in the country, Ayatollah Khomeini, the leader of the Iranian Revolution, in a message to the nation, referred to the necessity to transform the cultural and educational system of the country, thereby emphasizing the need for the creation of an organization that would plan different academic programs and the formulation of an academic strategy that addressed the future of the universities. In February 1981, the Iranian Cultural Revolution was proclaimed, and after the Revolutionary Council approved the terms of the Revolution, it subsequently declared the closure of all universities and institutions of higher learning beginning in the middle of June 1981.[2]

For full implementation of the Cultural Revolution, Ayatollah Khomeini proclaimed the formation of the Cultural Revolution Initiative in late 1981. The proclamation emphasized the achievement of important principles, including the need to bring meaningful, fundamental changes to all universities in the country: transformation of the universities into a healthy environment for the development of higher Islamic studies; the revision of academic programs and the training of instructors; the appointment of suitable instructional staff; the admission of the students; and conversion of the cultural environment of the universities to an Islamic one.

On the research and development front, due to eight years of national defense against the Iraqi invasion of the country, which drew the productive capacity of the country toward production for defense efforts, little progress took place in higher education, and only 27 new research centers were established. Accordingly, during this period, the key indicators of research and technology were in sharp decline in Iran. The number of researchers per million population fell from 38 to 28. The number of patents plummeted from 2,054 to only 179. The number of published papers in Institute for Scientific Information (ISI)-ranked journals had a free fall from 450 to 158. The R&D intensity, that is the R&D investment as a percentage of the gross domestic product (GDP), was reduced from 0.27 percent to 0.23 percent.

In summary, we point to the shift of ownership of many private business entities to government ownership, the exit of American experts from the Iranian military and industrial sectors, the occurrence of the Cultural Revolution, the imposition of a moratorium on higher educational activities, and the invasion of Iran by Iraq in the immediate aftermath of the Islamic Revolution in describing the prevailing conditions in Iran in the 1980s. Furthermore, during this period, the revolutionary upheaval, political instability, and the war caused social instability, disruption in management systems of the country, and a radical dilution of the quality of education and research in Iran. Additionally, overcentralization and the development of parallel higher education decision-making centers were among the difficulties facing the educational system during this time.

At this time, the medical education system came to be separated from the higher educational system and contributed to the rise in multidecision-making centers, even though the separation contributed to the unity of theory and practice, and fundamentally enhanced research activities in the healthcare delivery systems. The upward trend in the participation of the private sector in higher education and the adoption of nontraditional methods to deliver instruction, such as distance learning, which had emerged, earlier were interrupted. Of course, the establishment of Islamic Azad University as a not-for-profit institution, a unique institution of higher learning of that type in Iran, in the midst of a totally state-dominated higher educational system, also occurred during this time. The founding of Islamic Azad University did not reflect the intent of privatization on the part of the government, but was a market response to an ever-increasing demand for higher education. However, the success of this institution opened the way for private-sector participation in the higher education system in Iran.

In the first decade after the triumph of the Revolution and after the reopening of the universities, the development of higher education became the country's most important component of science policy. Given that the most important task of the manufacturing sector, which was mostly publicly owned, was meeting the war and private consumption needs of the country, the technology policy found meaning insofar as it could meet the daily requirements of the industry. In short, during this decade there was no systematic technology policy; however, there was some evidence of an isolated government industrial policy during the first decade after the Revolution. For instance, automotive industrial policy during this time aimed to have the car manufacturers develop domestic technological capabilities, and demonstrate self-reliance and establish belief in the country's own abilities, among other goals. Automotive industrial policy at the time was adopted according to which of the assembly plants received government attention and supports to transform themselves from plants for mere assembly of foreign manufactured auto parts to firms that could design and manufacture automotives. But despite the heavy support of government and even domestic design of some cars, Iran's car manufacturing industry couldn't become a leader in the global automobile industry and be competitive with other famous car producers.[3]

The Second Decade after the Triumph of the Islamic Revolution (the 1990s)

Iran's national research system was resuscitated by the Supreme Council of the Cultural Revolution in 1990 by a decree that defined the conditions for the establishment of research centers. This decree resulted in the coordination of affairs relating to the founding of research institutions and increased the number of public and private research centers. In 1988, 79 public research centers existed in the country, and in the next 10 years this number increased to 154, or a rise of roughly 100 percent in 1997. In this period, the number of researchers per million population had a remarkable growth from 45 researchers in 1990 to 333 in 1997. The number of ISI publications rose from 173 to 1,103 during the same time also. The number of patents increased from 299 to 315 in this decade. However, the rate of R&D expenditure out of GDP remained more or less constant at 0.33 percent.

After the end of the war, the government placed the opening up of international relations and the reorganization of government ministries and the state-owned enterprises, as well as the creation of conditions for rapid economic growth on its policy agenda. During this period, STI policy was not the center of attention; however, alongside the changes in the socioeconomic conditions, the acquisition and development of the sciences expanded in the country, and technology, which was transferred from abroad in many cases, and did not enjoy indigenous growth, gradually acquired relevancy for economic developmental policy makers.

In short, one could claim that during this decade the quantitative improvement in higher education played an important role in the country's science policy. Moreover, the policy of increasing graduate educational opportunities, especially during the closing years of the decade, and the expansion of research centers were the early signs of a changing science policy in Iran. The technology policy during this period played a subordinate role to industrial policy, and technology transfer from abroad, especially from Russia, Germany, and France played an important role too. Heavy and basic industries such as steel, aluminum, and concrete, mostly due to the national reconstruction needs after the war, were rapidly established in various locations. Additionally, important technological changes such as in the information and communication technologies, and the economic significance of emerging technologies were harbingers of the serious transformation of the country's technological policies, and sparked serious attention to the emerging technologies toward the end of the decade.

The Third Decade after the Triumph of the Revolution (the 2000s)

In this decade, a major transformation in the higher education system, mostly in the context of the Third Socioeconomic and Cultural Development Plan, took place, and it is considered a turning point in the history of higher education in Iran. With approval of the Third Development Plan by the lawmakers, and for the purpose of cohesiveness in policy making and the implementation of the goals of the educational system, the name of the Ministry of Culture and Higher Education was changed to Ministry of Science, Research and Technology, and the tasks of planning, support, evaluation, monitoring, and development of research and technological policies and priorities were added to the responsibilities of the newly formed ministry (Entezari 2009). This organizational shift corrected the tendency toward the decentralization of educational policy making, which in previous years had appeared in the form of the Ministry delegating the responsibility for curriculum planning and recruitment of faculty members to some of the universities.

A review of the top national policy documents, such as Third and Fourth Economic Development Plans, Vision 1404, and the law that established the Ministry of Science, Research and Technology, reveals developmental policies relating to science, technology, and to a certain extent, innovation.

The Third Socioeconomic Development Plan adopted some supporting policies for the development of research. These measures included the establishment of nongovernmental funds for research and technology, with assistance from the government and publicly owned banks, and the government's financial support of the cost of all specialized research projects with a ready-made demand, provided that the entity demanding the research contributed at least 40 percent of the cost of the project. This trend continued in the Fourth Socioeconomic Development Plan (2005–2009), and for the first time in the history of development planning in the country, an independent chapter that dealt with knowledge-based development was included in the plan. Six articles of this chapter pertained to issues related to the promotion of research and technology that the government is obliged to perform. Moreover, the articles referred to the implementation of certain actions such as the modernization and restructuring of strategies and plans for research, technology, and education for the country. These measures were designed to increase Iran's capabilities in science, technology, and its educational centers to meet the demands of industry and the society as a whole, within the framework of increasing global competition. Moreover, the government was required to devise plans for scientific and technological development according to global standards, support investment for production, provide technological information in the Farsi language, and achieve all of the above by reliance on the private and cooperative sectors of the economy

The latest macrodevelopment plan, the Fifth Socioeconomic Development Plan (2011–2015), contains a chapter dealing with the development of science and technology. This plan places special emphasis on the fundamental transition in higher education affairs, especially in the humanities, and accentuates the goal of Iran's acquisition of a second-place ranking in science and technology in the Middle East by 2015. This program also recommends planning to increase the annual expenditure on R&D by 0.5 percent of the GDP, so that by the end of 2015 it reaches 3 percent of the gross national output. Furthermore, the plan requires qualitative and quantitative improvements in the universities and the other institutions of higher learning; the expansion of scientific relationships with reputable international learning and research entities through the establishment of joint universities; the setting up of joint academic and research programs with foreign universities; and international exchanges of professors and students, with emphasis on the countries in the region and the Islamic world, particularly in the fields of humanities, religious studies, and the advanced sciences that have priority for the Islamic Republic (IR) of Iran. In the meantime, the plan emphasizes achieving these goals by creating harmony among the educational institutions in terms of the policy making, planning, and monitoring of the developmental policies by the Supreme Science Research and Technology Council, and the collaboration of the Statistical Center of Iran for assessment of the scientific conditions of the country on national, regional, and global scales. The plan mandated that the government establish assistance programs for the technological development and innovation in the private sector.[4]

A review of the science and technology (S&T) indicators in this period shows noticeable improvement in the developments in research and technology in Iran. For example, the number of researchers per one million people increased from 336 in 2000 to 664 in 2006. From the start of the Third Socioeconomic Development Plan in 2000, the number of innovation increased very rapidly. The number of patents by resident Iranians skyrocketed from 241 in 2000 to 10,118 in 2008. However, the patents have dropped precipitously since 2008 to around 6,000 in 2009, and fell to a little over 5,500 in 2010. The number of articles published in scientific journals also shows a remarkable improvement in terms of the country's scientific development. For instance, in 2000, Iranian resident researchers published 1,318 articles in ISI-ranked journals, while this number rose to 9,061articles in 2007, or a 6.9-fold increase in seven years. According to the statistics, Iranian scholars have published 21,038 articles, and with a 1.44 percent share of the world production of articles, Iran was ranked twentieth in research output in the world in 2011 (Hariri and Salahi 2012).

It seems that a significant portion of the key indicators of research and technology development in recent decades is due to the growth

Table 3.1 Key Indicators of Research and Technology Development in Iran, 2005–2009

Indicators	2005	2006	2007	2008	2009
Number of incubators	42	43	56	61	67
Number of S&T parks	14	14	19	22	25
Number of settled firms in S&T parks and incubators	674	824	979	1600	1982
Number of full-time knowledge workers who work in established firms in S&T parks and incubators	3,292	4,068	8,991	11,451	11,952
Number of patents granted to settled firms in S&T parks and incubators	82	145	236	245	449

Source: Institute for Research and Planning in Higher Education, 2011.

and expansion of research and technology infrastructure in this period. Table 3.1 shows phenomenal growth in these indicators. The number of science and technology parks increased by over 100 percent between 2004 and 2008, while the number of institutions located at the parks had a three-fold rise from 676 to 1,982 during the same time. We observe the same pattern of rapid expansion in the number of workers in these parks from 3,292 to 11,952, and the number of patents, which increased a little over five times during the same period.

Similar to the noticeable technological progress occurring globally during the first decade of the twentieth century, one could consider the third decade after the Iranian Revolution as a decade of rapid change and major transformation in science, technology and innovation. It could be said that the first part of the decade represented the golden years in terms of Iranian attention to the development of advanced technologies and technology policy. Among the most important technology policies of this period, we can cite the following measures:

- The activities of the President's Office of Technology Cooperation increased, and this office paid particular attention to emerging technologies, particularly nanotechnology. This office, which is known as the President's Office of Technology Cooperation, has played a role in step-by-step formation of the national system of the innovation or sectoral system of the country. The Center, which is one of the most stable active-technology policy making centers in the country, has contributed to the establishment of a network of activities or institutions that are part of the system of innovation, among which we include investment funds in knowledge-base activities and formation of the Pardis Science and Technology Park.

- During the early years of the first decade of the twenty-first century, the Ministry of Industry and Mining, through its Center of High-Tech Industries, invested in several large projects at the national level to support high-tech industries. The investments by the Center supported research projects, produced prototypes, and funded activities for the commercialization of inventions in new materials, nanotechnology, biotechnology, aerospace, laser and optics, software for information and communication technology, and new sources of energy. These financial supports resulted in the creation of technological companies in the private sector. We cite the investment in the bio-implants production project, which began with the financial support of the center, and resulted in the formation of Baft and Nosoj Zendeh Company, which succeeded in the industrial production of a bio-implant for human eyes for the first time in the world. Moreover, the studies relating to technology policy, which young researchers conduct as part of a think tank at the Center, and which emerged spontaneously and were independent of the activities relating to the Center's mission, were responsible for the formation of a new generation of research projects and the training of new researchers. However, due to political changes and repeated organizational restructuring in the Ministry of Mining and Industry, the activities of the Center have faced serious challenges and have not been able to achieve their previous standing.
- Other institutions such as the defense industry, the Power Research Institute, and the Research Institute of the Petroleum Industry (RIPI), because of the sanctions on the purchase of double-use technologies abroad, have devoted special attention to technological development in recent years. Because of the impediment created by the sanctions, many formerly imported technologies were developed domestically with the help of Iranian researchers and technicians.
- Starting in 2000, programs on technology management and technological policy making were introduced as undergraduate and postgraduate academic disciplines in Iranian universities. Currently, many of the alumni of these programs are serving in their respective fields in various industries and government agencies.

Along with these developments, the growing population and the increasing demand for higher education, the huge government investment in higher education, and the demand pull for advanced research in emerging technologies have created extraordinary demand for undergraduate and graduate education, especially for graduate studies. Accordingly, science policy during this decade is noticeably different than what it was in the past. In this period, dependence on foreign universities diminished, and the number of students who were sent abroad for education decreased. The educational philosophy of emphasis on simple teaching universities was changed to a policy emphasizing research universities, where the professors are required both to produce and disseminate knowledge. The rapid increase in the output of the educational system, like the rapid growth of published articles in international scientific journals, is telling evidence of this transformation. Progress in the technical expertise of Iranians working in advanced technologies such as nuclear, aerospace, and nanotechnology is the result of mission-oriented, wholly supply-side government technology policies, however.

The issue of the commercialization of inventions resulting from R&D, which arose in the West three decades earlier, also emerged in Iran during the 2000s. Focusing on innovation, taking a system perspective toward innovation, and stimulating innovation, appeared in policy documents toward the middle of the last decade, and it became a policy objective to develop a knowledge-based economy. However, there were no optimistic signs pointing to such achievements until 2010. Perhaps a lack of consistency in policy making, a lack of attention to the technological conditions in the country, the superficial replication of foreign experiences, and a lack of careful technology planning are some reasons for the difficulties in the national systems of innovation.[5] Some of the remedies for this ill include avoiding the establishment of parallel programs and eliminating carelessness in policy making.

Policy Documents on Science, Research, and Technology

As stated in the previous section, during the last 30 years many documents dealing with various science and technology policies were prepared in Iran. Table 3.2 shows a list of these documents and the governmental bodies that developed them. One could come up with two interesting observations by reviewing this table. First, some of these policies never were implemented, and second, some of the documents were replicas of the existing policies and documents. Given these difficulties, under the guidance and tutelage of the Supreme Leader, Ayatollah Khamenei, the Comprehensive Scientific Roadmap of Iran, as the most modern STI plan for the country, was prepared in 2010. This document was supplied to all governmental executive bodies in the country as the guiding principle of technological development, and it sketches the desired state of scientific and technological development for the country over the next 25 years.[6]

Table 3.2 Policy Documents on Science, Research, and Technology Formulated after the Revolution in Iran

Title of Policy Document	Year	Taskmaster	Policy Formulator Body	Authorized Body	Stakeholders	Enacted Policy	Enforced Policy
Vision of 1404	2002	Islamic Republic of Iran Expediency Council	Commissions of Islamic Republic of Iran Expediency Council	Supreme Leader	All the people and institutions in Iran	√	√
Strategy for Secure Exchange of Information in Iran	2004	Supreme Council for Secure Exchange of Information	Secretary of Council for Secure Exchange of Information				
Future Strategy: Ten years strategy for development of nanotechnology in Iran	2004	Iran Nanotechnology Initiative Council	Iran Nanotechnology Initiative Council	Cabinet and Supreme Council for the Cultural Revolution	All the public institutions involved in S&T development in Iran	√	√
Research and Technology Development Plan	2005	Management & Planning Organization	Management & Planning Organization – Ministry of Science, Research & Technology				
Strategic Industrial Development Plan of Iran	2005	Ministry of Industry & Mines	Sharif University- Faculty of Management & Economics				
Short-Term Strategic Plan for Development of Science and Technology in Iran	2007	Ministry of Science, Research & Technology	National Research Institute for Science Policy				
National Strategy of Iran Fuel Cell	2007	Ministry of Energy	Ministry of Energy	Cabinet	All the public institutions involved in S&T development in Iran	√	√

Continued

Table 3.2 Continued

Title of Policy Document	Year	Taskmaster	Policy Formulator Body	Authorized Body	Stakeholders	Enacted Policy	Enforced Policy
Strategic plan for Development of Research and Technology in Oil Industry of Iran	2007	Permanent Commission of Supreme Council for Science, Research & Technology – Ministry of Oil	Deputy for Human Resources & Deputy for research & Technology of Ministry of Oil	Ministry of Oil	All the public institutions involved in S&T development in the oil sector		√
Comprehensive system for Information Technology of Iran (Strategic plan)	2007						√
Comprehensive Strategy for Development of Iran's Aerospace	2008	Supreme Council for the Cultural Revolution– Ministry of Science, Research & Technology	Aerospace Research Institute	Supreme Council for the Cultural Revolution			
Strategic Evolution of Science and Technology of Iran	2009	Supreme Council for the Cultural Revolution	Ministry of Science, Research & Technology	Minister for Science, Research & Technology		√	
Comprehensive Scientific Roadmap of Iran	2010	Supreme Council for the Cultural Revolution	Many ministries and public bodies	Supreme leader- Supreme Council for the Cultural Revolution		√	√

Source: Tabatabaeyan and et al, 2011.

The Prospects and Challenges Ahead

A review of the history of Iranian policy making in STI shows a progressive development over the last century. The establishment of a formal educational system in the country one hundred fifty years ago, the growth of the higher education system, and establishment of infrastructure for research have gradually contributed to the continual development of technology in the country. The scientific-technological growth has enabled Iran to acquire certain advanced technologies such as the launching of satellites and the production of new pharmaceutical compounds. However, the country has faced challenges and difficulties in achieving these capabilities. Below we discuss these challenges and point to promising growth and development in these fields.

S&T Policy Making in Iran and the Challenges Thereof

The Country Faces a Number of Challenges in S&T Policy Formulation

• **Fluctuation in Oil Revenues**

The country's overreliance on oil revenues is one of the most outstanding challenges facing its scientific-technological development. Historical analysis of public policy making in Iran clearly shows that these policies are particularly vulnerable to declines in oil revenues. Therefore, US-led sanctions against the Iranian oil industry and central bank could be regarded as the next major challenge. However, some countermeasures have already been anticipated in order to prevent or eliminate the negative impacts the sanctions may impose on Iran's S&T policies. The Iranian Supreme Leader, for instance, on March 2, 2012, launched a major campaign called National Production, Supporting Iranian Capital and Labor to emphasize the development of products and technology indigenously. Implementing this policy will increase national income and reduce the country's dependence on the revenues generated from petroleum exports.

• **Overlapping Policies and Parallel Policy-Making Bodies**

The Iranian Revolution totally changed the policy-making arena in the country. It led to the closing of some entities, the changing of the missions of others, and the formation of new institutions based on the revolutionary ideals, and without adequate consideration of the implications of these changes. These changes, over time, resulted in overlapping policies and parallel policy-making bodies. Although the government dealt with some of these deficiencies, several policy-making institutions with duplicative and overlapping missions are still active. For example, the mandate of the Supreme Council for Informatics largely overlaps with the work currently carried out by the Supreme Council for Information, and the role of the Ministry of Science, Research and Technology is very similar to that of the Vice President's Office for Science and Technology.

• **Detachment of S&T Policy and Macroeconomic Policy**
It is widely recognized that S&T policies affect and are deeply affected by industrial and economic policies. This implies that S&T policies must be carefully devised within specific contexts of a general socioeconomic development plan. In Iran, however, S&T policy has been detached from its largely industrial and economic context. Although fairly good research and technology policies have been in place, other institutional frameworks and regulations in the larger industrial and economic context have been missing that should support the development and commercialization of new products and services based on locally developed technologies.

• **Ineffective Policy Monitoring and Evaluation**
Evaluation constitutes a critical component of policy making, and serves the important functions of improving the quality of the policy and assessing whether the goals of the policy have been achieved. Ineffective policy monitoring and evaluation mechanisms in Iran's S&T arena are posing a persistent challenge. This, in turn, has contributed to the misallocation of resources and lack of accountability. Policy makers should seriously consider the setting up of such a framework for monitoring and evaluating the implemented policies.

• **Lack of a Comprehensive System for Data Gathering in Support of S&T Policy Making**
A major part of the effectiveness of adopted S&T policies depends on access to exact and transparent data related to the latest state of technological development of the country. Such data are imperative for S&T planning for further development. Unfortunately, due to the lack of a comprehensive system for collecting and preparing S&T data in Iran, most of the policies that are adopted are backed up with a weak data-support system. However, in recent years activities for the collection of S&T information have been undertaken, and the first phase of the initiative will produce results in 2012.

• **Ignoring the Demand Side for Science and Technology Policies**
Generating markets and creating demand for indigenous technologies is one of the key elements in the development of science and technology systems in many advanced industrialized countries. However,

the creation of demand for innovation has received little attention in Iran, and most of the STI policies in the country have a supply-side orientation.

• **Neglecting Cooperation with Domestic and International Experts in Technology during Formulation of National S&T Policies**
In the formulation of many S&T policies, policy makers who have no training and experience in technology policy analysis have failed to take advantage of the expertise of domestic and international S&T professionals.

Prospects for the Future of the S&T System in Iran

In closing, we observe that with the growing number of expert and trained personnel in S&T policy making, and the increasing skill and experience that policy makers have developed after working on several national policy-making projects, it is hoped that the national system of innovation in Iran will evolve at an accelerated pace. On the other hand, formulation of the Comprehensive Scientific Roadmap of Iran for the development of science and technology, and as a guidepost for actions by the major players involved in the S&T scene could be a hopeful sign for the creation of a suitable division of labor and specialization, which would inevitably lead to preventing the waste of national talent and resources. Finally, the passing of a law in the parliament of Iran to support knowledge-based companies, the establishment of the innovation and development fund, the creation of necessary incentives to encourage the private sector to invest in advanced technologies, as well as the commercialization of innovations would promote innovation in the country and expand exports of high technology products.

Notes

1. For details about this university, please see chapter 2 of this volume on the history of the sciences in Iran.
2. The Cultural Revolution refers to a collection of activities of the Islamic Republic, mostly during 1981–1988, that aimed to bring about a deep transformation of Iranian academic culture.
3. See chapter 10 on the automotive industry in this volume for more details.
4. We refer the reader to detailed discussions of these plans in chapter 4 on the national innovation system of Iran in this book.
5. For a detailed discussions on the inadequate innovation in Iranian enterprises, see the introduction and chapter 4 on the national innovation system in this book.

6. For a detailed discussion of the Vision 1404 Plan, see chapter 4 on the national innovation system in this book.

References

Entezari, Yaghob. 2009. "Sixty Years Higher Education, of Science, Research and Technology" (in Farsi). CITY: Institute for Research and Planning in Higher Education Publications.

Ferasatkhah, Maghsoud. 2000. "The History of the Evolution of the University in Iran" (in Farsi). CITY: Institute for Research and Planning in Higher Education Publications.

Hariri, Nadjla, Maryam Shekofteh, and Ali Salahi Yekta. 2012. "Co-Citation Scientific Maps: A Case Study of Medical Sciences in Iran." *Journal of Paramedical Sciences* 3 (1): 47–60.

"National Report on Higher Education, Research & Technology" (in Persian). 2011. CITY: Institute for Research and Planning in Higher Education Publications.

"Statistics of Higher Education of Iran from 1925 to 1976. 1976" (in Persian). CITY: Institute for Research and Planning in Higher Education Publications.

"The Statistics of Public Research Institutions in Iran" (in Persian). 1987. CITY: National Research Institute for Science Policy Publications.

Tabatabaeian, H., S. Farnoodi, and R. Naghizadeh, R. 2011. "The First Book of Policymaking" (in Persian). CITY: National Research Institute for Science Policy Publications.

CHAPTER 4

The National Innovation System of Iran: A Functional and Institutional Analysis

Abdol S. Soofi, Sepehr Ghazinoory, and Sanam Farnoodi

It is well known by now that innovation plays a pivotal role in sustained economic growth and a high standard of living. In spite of this important insight, the first-generation economic growth models (Solow 1956) treated technical change as an exogenous variable, that is, the model did not explain how technology affects the economic growth of an economy. Recognizing this deficiency in neoclassical growth theory, a number of researchers concentrated their attentions on the subject and studied the systems of innovation, which appear to be different in various countries.

Traditionally, in explaining the process of innovation, many authors concentrated on national systems of innovation, a framework that was first introduced by Freeman (1987, 1), who defined it as "the network of institutions in the public and private sectors whose activities and interactions initiate, import, modify, and diffuse new technologies." Some researchers, objecting to the level of aggregation at the national level, adopted a new approach to innovation studies that focuses on a less aggregated, sectoral, rather than a national system (Malerba 2002).

Recently, recognizing that innovations that lead to economic growth have negative externalities (e.g., environmental pollution, global warming), which impose social costs, some innovation researchers have proposed a new innovation system-study framework for a better understanding of the processes of sustainable technological change. In doing so, they have concentrated on the technological innovation system (TIS) approach for innovation studies. This approach is important because emerging technologies evolve within specific TIS. Ultimately, the TIS approach aims toward the innovation of products and processes that could reorient economic activities, and that in turn can lead to the sustainability of technoeconomic changes (Hekkert and Negro 2009).

Researchers using the system approach in the study of innovation realized that in general, systems have functions and interactions among the workings of each system, which create the system's dynamics. The earlier approaches to innovation system did not consider these functions or the interactions. However, these functions—where a system function is defined as "a contribution of a component or a set of components to a system's performance" (Johnson and Jacobsson 2000)—play a pivotal role in the evolution of systems, and in the specific case of an innovation system, perform crucial tasks related to innovation activities and results, which ultimately leads to economic growth.

Viewing innovation from this system perspective, Johnson and Jacobsson (2000) defined a set of functions for innovation systems. Hekkert and colleagues (2007) used these functions in mapping the activities of the innovation system, followed by Hekkert and Hegro (2009) then empirically validating these functions in the framework of technology innovation systems.

These functions include:

F1. entrepreneurial activity
F2. knowledge development
F3. knowledge diffusion
F4. guidance of the search
F5. market creation
F6. resource mobilization
F7. creation of legitimacy

We will define these interactive functions in the following sections of the chapter below.

In this chapter, we use the system function framework to study the national innovation system of Iran.

System Functions of the Iranian National System of Innovation

With a land mass of 1,648,000 sq KM and a population of roughly seventy-six million in 2012, Iran had a gross domestic product (GDP) per capita of 12,258 at Purchasing Power Parity (PPP)[1] and in constant 2000 prices in 2011 (Wikipedia).

With these introductory remarks about Iran, we next turn to a functional analysis of the national system of innovation of the country.

Function 1: Entrepreneurial Activities

Before we discuss the entrepreneurial activities in Iran, it is instructive to define the terminologies we use in this chapter. First, we define innovation. According to the Organization for Economic Cooperation and Development (OECD) innovation is "the implementation of a new or significantly improved product (good or service), or process, a new marketing method, or a new organizational method in business practices, workplace organization or external relations" (OECD 2005, para. 140).

Second, to avoid confusion between innovation capabilities and innovation activities, we present the definition of these concepts also. OECD (2005, para. 149) defines innovation activities as "all scientific, technological, organizational, financial and commercial steps which actually, or are intended to, lead to the implementation of innovations." Romijin and Albaladejo (2002, 1054) define innovation capabilities as "the skills and knowledge required to make independent adaptations and improvements to existing technologies, and ultimately to create entirely new technologies."

Finally, the notions of entrepreneurship and innovation activities are confused occasionally. To differentiate between entrepreneurship and its relationship with innovation capabilities in an economy, we provide a definition of the former as well. However, entrepreneurship is an elusive concept to define, and there exists a vast literature that deals with defining the term (for example, see Radosevic 2005). We prefer to define an entrepreneur as a person who is willing and able to take advantage of entrepreneurial opportunities, whereas an entrepreneurial opportunity refers to "a situation in which a person can create a new means-ends framework for recombining resources that the entrepreneur believes will yield a profit" (Shane 2003, 18).

An inadequate level of innovation activities could have two causes: lack of adequate entrepreneurship or a weakness in the innovation system (Bascavuoglu-Moreau 2010). Therefore, it is important that one distinguish between a lack of entrepreneurship and an inadequate innovation system in an economy. In this section, we rely on the Global Entrepreneurship Monitor's (GEM's) study of entrepreneurship across countries to assess the degree of entrepreneurship in the Iranian economy, and examine the innovation system of the country in other sections.

The World Economic Forum classifies economies according to their developmental stages: Factor-driven, efficiency-driven, and innovation-driven. The factor-driven development phase is characterized by mostly subsistence agriculture, labor-intensive methods of production, and natural-resource extraction. The efficiency phase of development is dominated by the large-scale industrial organizations that often enjoy economies of scale and capital-intensive production methods. Lastly, in the innovation-driven development phase, firms adopt knowledge-intensive methods of production, and the economy has a large service sector (Schwab 2010).

Using this classification scheme, the GEM places Iran in the factor-driven phase of economic development, even though the Report indicates that Iran is in transition to the next phase, that is, efficiency-driven (Kelley, Bosma, and Amoros 2011).

According to the GEM model, the prerequisites to innovation-driven economic growth consist of three categories of variables: the basic requirements (institutions, infrastructure, macroeconomic stability, health, and primary education); the efficiency enhancers (higher education and training, goods market efficiency, labor market efficiency, financial market sophistication, technological readiness, and market size), and innovation and entrepreneurship (government policy, government entrepreneurship programs, entrepreneurship education, research and development [R&D] transfer, internal market openness, physical infrastructure for entrepreneurship, commercial, legal infrastructure for entrepreneurship, cultural and social norms) (Kelley et al. 2011). Unless a country meets the prerequisites, the authors of the Report contend that entrepreneurial finance (venture capital) and government investment in entrepreneurial programs are less likely to succeed since the basic requirements are not present in the economy.

We will examine the entrepreneurial activities in Iran within this analytical framework.[2]

1. Entrepreneurial attitude and perception

We present the results of a 2010 survey of entrepreneurial attitudes and perceptions of the GEM in Table 4.1. The survey measures the general public's attitudes toward and perceptions of entrepreneurship. A positive public attitude creates beneficial externalities such as cultural

Table 4.1 Entrepreneurial Attitudes and Perceptions: Select Countries

	Perceived Opportunities	Perceived Capabilities	Fear of Failure	Entrepreneurship as a Good Career Choice	High Status to Successful Entrepreneurs	Media Attention for Entrepreneurship	Entrepreneurial Intentions
Iran	41.6	65.7	30.1	63.6	84.6	62.3	31.4
Turkey	36.1	54.2	25.0	71.2	76.4	61.7	19.4
Germany	28.5	41.6	33.7	53.1	77.1	49.0	6.4
China	36.2	42.3	32.0	70.0	76.9	77.0	26.9
USA	34.8	59.5	26.7	65.4	75.9	67.8	7.7
Variance	23.83	118.11	10.53	41.51	17.35	83.40	128.49

Source: GEM 2010 Global Report.

supports, financial resources, and networking opportunities for current and future entrepreneurs. The questionnaire involved a number of categories that appear as the headings of the columns in Table 4.1.

In addition to Iran, we include a representative sample of countries from the Middle East (Turkey, efficiency driven), the Asia-Pacific (China, efficiency driven), Europe (Germany, innovation driven), and the Americas (the United States, innovation driven) in the table for comparison. The choice of sample countries is based on our observations that the United States and Germany have the most advanced National System of Innovation (NSI) in the Americas and Europe, respectively; China has the most dynamic, robust economy in Asia and the world; and Turkey is a neighboring Middle Eastern country of Iran. Because of cultural similarity, geographical vicinity, and stage of economic development shared by Iran and Turkey, in discussion of some of the questions in Table 4.1, we compare Iran and Turkey only.

The numbers in Table 4.1 are percentages of the total respondents who expressed particular attitudes. For example, 41.6 percent of the respondents in Iran perceived good opportunities for entrepreneurship. The percent of the respondents who perceived good opportunities in Turkey is 36.1 percent, in Germany 28.5 percent, in China 36.2 percent, and in the United States 34.8 percent.

Next, almost two-thirds of the Iranian respondents perceived confidence in their own entrepreneurial capabilities. The figure drops to a little over 54 percent for Turkey, and to a low of 41.6 percent in Germany.

The third question of the survey involved assessment of fear of failure. A relatively large percentage of Iranian respondents showed fear of failure (30.1 percent) compared to the Turkish respondents (25 percent), and the US respondents (26.7 percent). This finding may reflect the more volatile macroeconomic and governmental policy environments as well as the relative unreliability of the judiciary system in Iran compared to those in Turkey and the United States.

A larger percentage of Turkish respondents (71.2 percent) considered entrepreneurship a good career choice compared to the almost 64 percent of the Iranian respondents who believed entrepreneurship was a suitable career option.

Almost 85 percent of the Iranian respondents perceived successful entrepreneurs to enjoy high status in society compared to 76.4 percent of the Turkish respondents who believed the same.

Regarding the question of whether entrepreneurs received a high level of media attention, the proportions of the respondents in the

Iranian and Turkish samples are roughly equal: 62.3 percent for Iran and 61.7 percent for Turkey. This finding, in addition to the large proportion of respondents who believe that successful entrepreneurs enjoy high status implies that these societies consider successful entrepreneurship a highly prestigious profession.

The last category, which shows the intention of the respondent to start a business, is very high for Iran (31.45 percent) compared to Turkey (19.45 percent) or the other countries in the table. In fact, only China, with almost 27 percent of the respondents who intend to start a business, is closer in value for this category to Iran. The high values of respondents who intend to start a business in Iran and China may reflect the lower employment opportunities in these countries, which force many unemployed workers to start businesses.

Next, we turn to analysis of the motivation as well as longevity of entrepreneurship based on the GEM data. The GEM Global Report 2010 lists several motives for people who start a business. These include starting of a business out of necessity (by unemployed workers), the desire to maintain and increase income (in profit form), and the desire to be independent (Kelley et al. 2011). The scatter diagram of the Total Early-Stage Entrepreneurial Activity (TEA)[3] and the purchasing power parity adjusted per capital GDP of 59 participating countries in the 2010 survey shows a U-shape distribution. This implies that the TEA starts at a higher level for the countries with lower per-capita GDP, reaches a minimum for the middle-income countries, and then rises for the high-income, efficiency-driven economies.

Table 4.2 below shows some additional survey results for the listed countries. Comparing Iran with Turkey, we see that a higher percentage of the respondents in Iran (12.4 percent) were involved in TEA compared to 8.6 percent of the respondents in Turkey. The established business ownership rate in Iran (12.2 percent) is marginally higher than that in Turkey (10.7 percent), while the rate of discontinuation of business in Iran (7.3 percent) is considerably higher than that in Turkey (4.6 percent). The necessity-driven businesses out of the total TEA in both countries are roughly equal, with 38 percent and 37 percent in Iran and Turkey, respectively. However, the improvement-driven entrepreneurship in Turkey (47 percent) outweighs the same for Iran (39 percent).

Based on these observations, we conclude that entrepreneurial activities in Iran in comparison to these activities in the selected countries are robust. However, to a certain extent, these activities in the emerging economies of China, Iran, and Turkey are necessity driven, rather than improvement driven. As will be discussed in the functional sections

Table 4.2 Participation of Individuals in Entrepreneurial Activity

	Nascent Entrepreneurship Rate	New Business Ownership Rate	Total Early-Stage Entrepreneurship Activity (TEA)	Established Business Ownership Rate	Discontinuation of Businesses	Necessity-Driven (percent of TEA)	Improvement-Driven Opportunity (percent of TEA)
Iran	4.8	7.8	12.4	12.2	7.3	38	39
Turkey	3.7	5.1	8.6	10.7	4.6	37	47
China	4.6	10	14.4	13.8	5.6	42	34
Germany	2.5	1.8	4.2	5.7	1.5	26	48
USA	4.8	2.8	7.6	7.7	3.8	28	51
Variance	0.987	11.67	16.23	10.89	4.633	47.20	49.70

Source: GEM 2010 Global Report.

below, much of entrepreneurship in both Iran and Turkey, particularly in Iran, does not involve innovation in the sense defined above. Accordingly, we conclude that inadequate innovation in Iran and Turkey is primarily due to a weakness in the innovation system of the countries. This means that the technological/institutional systems of the countries have not adequately matured so that they can encourage and nurture innovations. We will elaborate this point in the discussion section of the chapter below.

Function 2: Knowledge Development

Clearly, without knowledge, no R&D activities are possible, and without R&D investment, no technological progress is feasible. Therefore, the starting point for any national system of innovation is knowledge creation, which takes place first through formal education. In this respect, the role of the state is paramount. Accordingly, we will discuss the role of the state in the development of science, technology, and innovation policies in Iran. Furthermore, we will examine the level of educational achievements in the country.

The Iranian Education, Research, and Technology Policy-Making System

Formal university education in Iran began 1,700 years ago when Ardeshir Papagan (180–241 AD), the first Emperor of the Sassanid dynasty, commenced formation of Jondishapour University.[4] However, the first formal governmental institution of higher learning in modern Iran, Dar-ol-foonon, was established in 1852. Dar-ol-foonon, designed to be a polytechnic university, was modeled on the educational experiences of Russia and Ottoman Turkey, countries with more advanced technologies compared to Iran, and was the predecessor to the University of Tehran that was established in 1934. Moreover, prior to the establishment of the University of Tehran, schools of diplomacy (1898), medicine (1903), law (1910), veterinary medicine (1912), cultivation and farming (1922), commerce (1926), and the American School of Nursing, also known as Mason Nursing School (1926), were established (Institute of Research and Planning for Higher Learning [IRPHL] 2010).[5]

Given the number and variety of these learning institutions, the need for an organized governmental body to oversee the educational needs of the country became apparent, and the Ministry of Education was established in September 1909. Some of the duties of this ministry included the administration and auditing of elementary (free and mandatory), middle, and high schools as well as higher education, and the selection of high school graduates for studies abroad at the government's expense.

As the number of universities and other institutions of higher learning in the country increased, the Ministry of Sciences and Higher Education was established in 1967.

Shortly after (less than a month) the triumph of the Islamic Revolution in Iran (February 11, 1979), the new revolutionary government adopted a new educational paradigm and merged the former Ministry of Culture and Arts, and the Ministry of Science and Higher Education into the Ministry of Culture and Higher Education. Formation of the new ministry was the first step in a major transformation of the science and education philosophy in postrevolutionary Iran.

The Iranian universities were the centers of postrevolutionary upheavals in the country, and to address the problem of the revolutionary zeal of the students, which had detrimental effects on the educational processes and governance of the country, the revolutionary government first created the Cultural Revolutionary Council on May 23, 1981, and three years later formed the Supreme Council of the Cultural Revolution on December 10, 1984. During this transition period, all universities in the country were closed for the two academic years of 1980–1981 and 1981–1982. The universities and other institutions of higher learning resumed their operations for the second half of the academic year 1982–1983.

In 1983, the government separated the Iranian medical colleges from the main body of the country's higher educational system and made the management of medical training a function of a newly created Ministry of Health and Medical Education. In 2000, the government proposed a piece of legislation to the Parliament of the Islamic Republic (IR) of Iran that contained a plan to add to the duties of the Ministries of Culture and Higher Education the goals of "planning, support, evaluation, observation, examination, and writing policies, defining priorities in research and technology." It further proposed a name change for the Ministry of Culture and Higher Education to the Ministry of Science, Research, and Technology (MSRT) (IRPHL 2010, 14).

Finally, in 2004, the parliament passed a law that assigned additional tasks to the MRST. These additional goals consist of "developing science, research and technology; strengthening the spirit of research, creativity, and scientific culture; improving the educational, scientific, and technological conditions of the country; developing the Islamic and humane values; promoting arts, aesthetics, and the Iranian-Islamic scientific and literary tradition; securing expert labor; developing human resources; raising the level of knowledge and technical expertise; promoting the culture of science in the society; achieving academic freedom and the independence of the universities and centers for research and development; developing practical plans for higher education, science, research, and technology; supplying plans to the Council of Sciences; conducting R&D; identifying the comparative advantages, capabilities, talents, and R&D needs of the country; evaluating the priorities of research and technology by cooperation with or suggestions to the relevant executing organizations and suggestions to the Council of Science, Research and Technology; supporting development of basic research and projects related to the emerging technologies based on national priorities; controlling the processes of technological transfers and technical knowledge for the purpose of domesticating imported technologies; and adopting suitable policies and practices for the development of R&D in the private sector" (IRPHL 2010, 14).

The higher education, research, and technology decision-making bodies in Iran may be classified into three levels: national, sectoral, and university-wide. We will discuss these decision-making bodies below.

- The decision-making bodies at the national level

The decision-making bodies at the national level consist of the following institutions:

1. The Supreme Leader
    The Supreme Leader of the IR of Iran is the highest political authority in the country, who formulates the strategic vision for politico-economic development (including science and technology [S&T] policies) and the foreign policy of the Republic.
2. The Expediency Discernment Council of the System
    This body has the consultancy role for the Supreme Leader in the formulation of a general strategic vision for the country. Moreover, this body has the responsibilities of observation and coordination of the policies, as well as evaluation of the outcomes of the implemented policies. The Expediency Council plays an important role in S&T policy making in the country at present.
3. The Parliament of the IR of Iran
    According to the Iranian Constitution, the parliament can pass laws within the limits of its constitutional authority. Accordingly, the Education and Research Commission of the parliament of the

IR of Iran has the responsibility of examining and legislating laws pertaining to higher education, research, and technology.

4. The Supreme Council of the Cultural Revolution

   The Supreme Council is responsible for overseeing and designing educational, research, and technology policies within the general framework of the Islamic regime (power structure). The main goal of this organization is guidance of the educational system of the country so that it does not deviate from the Islamic principles and doctrines. Furthermore, it aims to develop an educational system that works against any intrusions of foreign powers to convert the educational system into a tool that could be used for the elimination of the national independence. Among the other responsibilities of this decision-making body is approving recommended individuals to be presidents of the universities. The resolutions of this body do not require approval by the parliament of the IR of Iran, and become laws automatically. As such, this organ is the most important decision maker in the formulation of S&T policies in the country presently.

   The Supreme Council examines and evaluates the scientific and research activities in the country annually.

5. The Supreme Council of Science, Research, and Technology

   The Supreme Council is chaired by the President of the Republic. The main responsibilities of the Council are (1) prioritizing major long-term educational, research, and technological investment projects, (2) suggesting and examining the sources of required financial resources for investment in higher education, research, and technology.

6. Deputy President for Science and Technology Affairs

   This organ is a newly created body in Iranian S&T policy-making establishment, and has assumed an increasingly important role in recent years. The importance of this organ is due to presence of the Office of Scientific Cooperation of the President and Office of Management of Talents in this decision-making unit. Moreover, this Office has assumed the responsibility of allocating and awarding a major portion of the government's annual budget for research in the country.

7. President's Cabinet

   The main functions of the executive branch in the education and technology decision-making processes are the allocation of funds for these purposes in the annual budget, and the issuance of directives and bylaws for the organizations that are responsible for education, research, and technology development. Moreover, the following ministries play important roles in the design and implementation of technology policies: the Ministry of Science, Research and Technology (MSRT); the Ministry of Industry, Mines and Commerce (MIMC); the Ministry Of Agricultural Jahad; and the Ministry of Health, Treatment and Medical Education (MOH). Moreover, the Technology and Innovation Cooperation Center (TICC) plays an important advisory role to the president of the Republic. Another important component of the President's Cabinet in S&T policy decision making is the Office of the Deputy President for Technical Affairs. This office has the responsibility for developing the 5-year socioeconomic development plan for the country. S&T development planning is part of the overall developmental plan, and in the absence of a general science and technology policy plan for the country, the plans offered by this office play an important role in the S&T policy development of the country. Since innovation and technological development mostly occur in the manufacturing and mining sectors of the economy, the Ministry of Industry, Mining, and Commerce has a pivotal role in advancing technological development and innovation. However, in spite of its important function, this ministry has played little or no role in guiding the innovations in the country (IRPHL 2009).

- The decision-making bodies at the sectoral level

Higher education, research, and technology policy making at the sectoral level is in the domain of the Ministry of Science, Research, and Technology and the Ministry of Health and Medical Education. The decisions are made at different councils in these ministries, including the Council of Development of Higher Education, the Supreme Council of Planning, the Council of Medical Education and Specialization, the Central Council of Scholarship, the Council of Guidance of Talented Students, the Council of Inspection and Evaluation of Higher Education, and the Council of Scientific Categories.

- Decision making and management at the university and institutions of higher learning

The board of regents, the university presidents, and university councils are responsible for education, research, and technology policy decisions at this level.

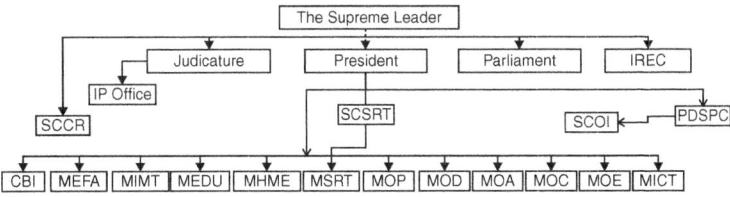

IREC: Islamic Republic of Iran Expediency Council
SCSRT: Supreme Council for Science, Research, and Technology
SCCR: Supreme Council for the Cultural Revolution
CBI: Central Bank of Iran
MEFA: Ministry of Economy and Fiscal Affairs
MEDU: Ministry of Education
MHME: Ministry of Health and Medical Education
SCOI: Statistical Center of Iran
MSRT: Ministry of Science, Research, and Technology
MOP: Ministry of Petroleum
MOD: Ministry of Defense
MOA: Ministry of Agriculture
MIMT: Ministry of Industry, Mine, and Trade
MOE: Ministry of Energy
PDSPC: President Deputy of Strategic Planning and Council
MICT: Ministry of Information and Communication Technology

Figure 4.1 Governance of Science, Technology, and Innovation in Iran

Based on the discussions of the decision-making bodies, Figure 4.1 shows a picture of the entities that are involved in the processes of policy planning and implementation pertaining to education, research, and technology in Iran.

Educational Attainments

The educational attainments of the country are shown in Table 4.3.

Table 4.3 indicates the state's educational expenditure as a percentage of the GDP and the general level of educational achievement for the country. Overall, the Iranian government tends to spend between 4 to 5.5 percent of the GDP on education. The Iranian investment in education has resulted in almost 26 percent growth in the adult literacy rate during almost two decades from 1990 to 2008

We consider as well other indicators of knowledge production and knowledge base in Iran.

The number of students enrolled in tertiary education, which means students who are enrolled in postsecondary institutions, has increased drastically over the last decade. The total enrollment in tertiary education in Iran was less than a million. This figure has leaped to over 4.5 million in 2011. This rapid increase in enrollment in colleges and universities was based on a deliberate governmental policy to raise the number of college graduates in the country (Mehr News, Report on numbers of students, 2011).

Table 4.3 Education Statistics

Year	Adult literacy rate (both sexes) (percent aged 15 and above)	Expected years of schooling (of children)	Expenditure on education (percent of GDP) (percent)	Mean years of schooling (of adults)
1990	65.5	9.7	4.1	3.7
2000	77.0	12.2	4.4	5.1
2005	82.4	13	4.7	6.1
2006	82.3	13.7	5.1	6.4
2007	82.3	14	5.5	6.6
2008	82.3	14	4.8	6.8

Source: UNDP, International Human Development Indicators.

Comparing knowledge creation and scientific output in Iran, measured in terms of the number of published papers, illustrates the efficacy of the country's educational system. The number of articles in all fields published in ISI-ranked journals[6] authored by scholars who resided in Iran at the end of the shah's rule in 1979 was around four hundred. This number increased to twenty-one thousand published scientific papers by 2011. Apparently, Iran has the highest rate of growth in publications in ISI-ranked journals in the world (Hamshahrionline 2011; Coghlan 2011).

R&D Outcomes

The number of funded research projects and the number of patents emerging from such investments are important indicators of the efficacy of investment in technological development. A total of 28,877 research projects, consisting of 11,025 government and 17,852 university-based research projects, were underway in 2009. As far as the number of patents is concerned, Iran has not been as successful as it has been in the publication of research papers and results. The country registered 7 patents with the United States Patent Office in 2010. Overall, between 1989 and 2010, Iran had 88 patents (Ghazinoory, Abdi, and Bagheri 2010).

Another indicator of positive returns on investment in S&T is the number of graduate degrees earned. The number of earned master's and doctoral degrees in 2009 was 24,248. Moreover, even though the number of full-time equivalent (FTE) researchers in Iran was only 82 per million of population, the number has drastically increased to 1,500 FTE per million in 2010 (Soofi and Ghazinoory 2011).

Function 3: Knowledge Diffusion through Networks

The Research Organizations

The main players in the Iran NSI are government ministries, universities, public and private research institutes, and large state-owned enterprises. Other players such as business associations, business-support organizations, consumer groups, and small and medium-size enterprises are very weak, and play little or no role in the innovation processes. A unique feature of the Iranian NIS is that all the major players are state owned. The Ministry of Science, Research, and Technology plays an important role in coordinating and evaluating the activities of the university research centers, public and private research institutes, technology parks, and business incubators of the country (IRPHL 2009). Nevertheless, in spite of the efforts of MSRT and the government ownership of these institutions, close links between these institutions rarely exist. However, these government-run institutions have well-developed knowledge systems and a reasonably well-developed manufacturing capacity in certain industrial sectors such as automobiles, telecommunications, and pharmaceuticals. Regrettably, the knowledge and manufacturing capabilities of the innovation system have not penetrated to small and medium-size enterprises and markets. The private sector firms have no significant role in the national system of innovation in Iran. Even the large state-owned enterprises confine their activities to the production of goods, and have little or no on-site R&D activities. Accordingly, the innovation system, apart from the government agencies and institutions, including the universities, has an underdeveloped network of knowledge diffusion (UNCTAD 2005).

The Iranian innovation system is a supply-push system. It is based on the implicit philosophy that "supply creates its own demand." That is, the implicit assumption of the government S&T decision-making establishment is that industry will learn to use a technology that has been supplied. Clearly, this system is totally different from a demand-pull system of innovation, where to meet the market demand and rigor of competition, enterprises are forced to invest considerable resources in R&D for innovation. We discuss this latter point in the historical context of the development of industrial research activities and innovations of large integrated enterprises of the West in the section below.

The supply-push system has established an elaborate network of research organizations in a variety of scientific fields in the country. To identify the variety of scientific fields of interest for the academic community and technology policy planners in Iran, we enumerate some of the most prominent centers and institutes below (IROST 2011):

1) Polymer and Petrochemical Research Center
2) Research Center for Fundamental Knowledge (Research Center for Theoretical Physics and Mathematics)
3) Research Center for Chemistry and Chemical Engineering
4) Petroleum Industry Research Center
5) Research Center for Social Sciences and Cultural Studies
6) Center for Information Science and Technology of Iran
7) Center for Strategic Studies
8) National Center for Research on Genetic Engineering and Biotechnology
9) Center for Research on Materials and Energy
10) Center for Research on Power
11) Center for Research on Aerospace
12) National Institute for Nuclear Energy of Iran
13) Institute for New Energies of Iran
14) Center for Commercial Information on Food and Agriculture of Iran
15) Center for Research on Biochemistry and Biophysics of the University of Tehran
16) Research Center for Building and Housing
17) Research Center for Science Policy of Iran
18) Center for Communication Studies of Iran
19) Center for International Energy Studies
20) Center for Biotechnology Studies of Iran
21) Iranian Statistical Center
22) Research Center for Protection of Soil and Watershed
23) Regional Center for Information of Sciences and Technologies
24) Institute for Standards and Industrial Research of Iran
25) Institute for Research on Agricultural Biotechnology of Iran

Conferences and Professional Meetings

One of the indicators of knowledge diffusion is the number of scientific and technical meetings and conferences held in a country. During the last two decades, the number of these professional meetings in Iran has increased dramatically. For example, 2,613 professional meetings were organized in the country during 2009. The government ministries and organizations organized 707 of these meetings, while the remaining 1,906 were organized by the universities.[7]

Networks
Another indicator of knowledge diffusion is the number and level of activities of the networks in a national innovation system. The main function of networks in innovation processes is the exchange of information among the participants in the networks. Since optimal S&T policies should be based on the knowledge of the existing technologies and the technological challenges faced by the society, the existence of the networks and communications among the networks is a prerequisite for learning and creating knowledge.

The idea of establishing scientific/technological networks in Iran emerged for the first time in 1991. The National Council for Scientific Research developed and implemented the first plan for establishing networks of research laboratories in biotechnology, information technology, seismology, and new materials science in 1997. The plan resulted in ten successful networks of the laboratories that were dispersed throughout the country. However, it appears that the established networks, which consist of government or university-owned research centers and institutes, have not created an environment that can rapidly disseminate knowledge to the industries, even though the networks are bound to play a useful role in exchanging knowledge among them. Another impediment in the efficient working of these networks is their dependency on these organizations as far as changes in the political leadership of the country. As government institutions, their functioning is affected by changes in the political leadership.

Scientific Journals
Yet another indicator of a system for the successful diffusion of knowledge in a NIS is the number of scientific journals. In Iran, the Ministry of Science, Research and Technology as well as the Ministry of Health, Treatment and Medical Education are the gatekeepers of the scientific integrity of journals. To receive accreditation, all journals must receive approval for publication from these ministries. The number of journals published inside Iran was in the neighborhood of 740, in March 2012.

Function 4: Guidance of Research

In 2004, the Iranian government developed a blueprint for the technological development of the country, a document that is known as

Vision1404 (Vision 2025 AD). The number 1404 in the title refers to year 1404 according to the Iranian calendar. This document specifies the strategic vision and goals of the scientific, technological, and economic development of the country over a 20-year period. Vision 1404 specifies the goal as: "The Islamic Republic of Iran in its Vision 1404 will be a capable country in the production and development of science, technology, and innovation; will be at the frontiers of the sciences; will be advanced in the utilization of able, learned, conscientious, and socially responsible human beings who have been schooled in Islamic thought who will be functioning at the level of the best scientists of the world; will achieve the first rank among the countries of Southwest Asia in science and technology; will be distinguished in some emerging technologies; and will have plans to generate wealth and national power" (The Expediency Discernment Council of the System 2003, translated from Farsi by the authors).

Vision 1404 was the first national overall S&T development plan in the country. However, as a general plan, the document did not enunciate specific goals and plans for actions to reach the goals. In 2010, another document, known as the General National Scientific Plan (The Supreme Council of Cultural Revolution 2010) was issued that defines the long- and medium-term objectives of the strategic goals of Vision 1404, in which the methods of reaching the goals in the areas of S&T were established.

Function 5: Market Formation

One of the government programs for market formation for the new technologies in Iran is the Maximum Utilization of the Domestic Production Capability Law, which was passed in December 1996. This law aims to support and promote domestic production of goods and services. According to the law, all government agencies and companies are required to purchase domestic engineering and technical consulting services, domestically produced goods, or those goods that the domestic companies have the ability to produce. Moreover, the law requires that at least 51 percent of any commodity used by Iranian government-owned firms or government agencies must be produced domestically. Additionally, this policy intends that the Innovation Development Fund's purchases of goods produced by newly organized innovative firms (spin-offs from universities and government research centers) should add to the creation of markets for innovations.

Many Iranian newly marketed products that were produced by the use of emerging technologies face numerous technical deficiencies, which makes marketing these products rather challenging. Facing these problems, many consumers prefer to buy older but reliable, well-functioning products. To overcome these demand-side obstacles by helping the producers to lower the prices of the new products, the Iranian government supports suppliers who produced by emerging technologies by providing fifteen-year tax write-offs and by promoting the export of these products through subsidies and the elimination of tariffs on imported capital and intermediate goods. These policies are in accord with the Maximum Utilization of the Domestic Production Capability Law of 1996.

Other government programs that promote the research and innovation of firms in the emerging technologies include low-interest loans, tax incentives such as exemptions amounting to 0.02 percent of total annual sales for companies that conduct research, and a $30 million fund for research, technology development, and innovation in newly established firms in the high-technology sectors. The fund has several responsibilities, which we enumerate below.

First, the new high-technology firms are supported by grants through the program. Moreover, the program provides long-term, subsidized loans to the well-established firms that aim to convert knowledge to wealth according to the national S&T policies. Second, the fund has the responsibility of assisting the country's venture capital industry or directly providing venture capital to promising new technology firms. Third, the fund insures there will be newly developed products and provides health insurance coverage for workers at the newly formed companies.

Finally, it is well known that macroeconomic policies and conditions are important determinants of business activities. Due to highly volatile macroeconomic policies, high inflation, and the unsettling unemployment problems in the country, many entrepreneurs do not find the macroeconomic conditions in the country conducive to business activities and risk taking. This factor may act as a countervailing force for further market formation in the country.

Function 6: Resource Mobilization

Next, we examine the resource mobilization for technological development by the government in Iran. Iran has an extensive university system. There are 119 government owned and operated universities, with 739 campuses that specialize in training students in science and technology. One of the government-owned universities, Payam Noor, has 550 campuses. There are 600 semipublic and private small universities throughout the country. The largest private university, Azad (Free) University, has 350 campuses. In the first quarter of 2012, the Ministry of Science, Research and Technology announced that the total number of these campuses had passed 2,400.

The government has allocated an increasing amount of investment funds to higher education development over the last several years. As a result, enrollments in doctoral programs totaled 38,910; in specialized doctoral programs 53,101; in master's programs 290,679; in bachelor's programs 2,680,817; and in the programs of technical colleges 1,530,086 during the 2011–2012 academic year (Mehr News 2011).

In addition to greater investment in education, the country has allocated additional funding for R&D. The government R&D expenditure as a percentage of the GDP ranged from 0.49 percent of the GDP in 2001, to 0.62 percent of the GDP in 2009.

Starting in 2010, the government mandated that 1 percent of funds allocated to the budgets of ministries and state-owned enterprises (SOEs) be spent on R&D according to the national priorities. However, based on personal observations of two of the authors of the present chapter, it appears that the ministries, agencies and SOEs were not successful in investing in R&D projects during the first year of the mandated rule. To assure the efficacy of the investments in R&D projects, the government established a project evaluation system to assess the outcome of investment projects that began in 2009.

Concluding this section, we note that the Iranian government has concentrated on supporting education more than supporting R&D over the last decade.

Function 7: Advocacy Coalition (Creation of Legitimacy)

The Supreme Leader, Ayatollah Khamenei, is the lead advocate for the technological development of the country. His main motivation for the advocacy role in this regard is his concern for the national security of the country. In a speech to Iranian professors in Tehran, he stated: "Colonial countries are quite aware of the fact that in order to keep a country under their political and economic dominance, they should bar its scientific progress" (Khamenei, 2005). This advocacy role of the Supreme Leader has been consistent throughout the years. In the latest of such public pronouncements in support of the development of S&T,

the Leader stated in a meeting of "Young Talented Individuals" in Tehran that, "Government officials must increasingly allocate funds for production of science and scientific innovation." Furthermore, the Supreme Leader emphasized the importance of the university-industry connection by stating that, "These Centers could meet the innovation needs of the industrial sector and could result in quality improvements of products, and at the same time the link could positively stimulate the scientific curiosity of the university researchers" (Tabnak October 5, 2011).

This advocacy for technological development by the Supreme Leader has resulted in focusing the attention of experts and government officials on the technological development of the country, particularly the development of emerging technologies.

Discussions

Based on the preceding discussions, we can infer that Iran, like many other emerging economies, suffers from major challenges in the industrialization and creation of a knowledge-based economy. We have observed that Iran has been highly successful in two of the functions of NIS, that is, knowledge creation and research guidance. The country has taken major strides in market formation also. However, the weakest point in the system is in transferring the acquired knowledge and technology to the industries where the entrepreneurial activities take place.

What factors contribute to this undesirable outcome in Iran? In answering this question, we rely on a recent opinion survey of policy makers and technology experts in Iran (IRPHL 2009), and a historical study of the pattern of the emergence of R&D departments in large integrated firms in the West and Japan.

Starting with the Iranian study, the MSRT conducted an opinion survey of influential individuals in Iranian S&T policy-making circles, and of active S&T experts who are engaged in planning for the national systems of innovation. The sample consisted of 115 respondents, 35 to 40 percent of whom completed the survey questionnaire. The respondents consisted of individuals from the Office of the Supreme Leader; the Ministry of Science, Research and Technology; the Ministry of Industry, Mining and Commerce; the Ministry of Defense; the Ministry of Information and Communication; the Ministry of Guidance; the Ministry of Power; the Ministry of Cooperatives; the universities; the Supreme Council of the Cultural Revolution: and other less prominent government organizations.

One of the questions in the questionnaire asked the respondents to characterize the existing style of governance that is part of S&T efforts based on the following list of management styles: command (hierarchical, from top to bottom); collective management (decisions based on interaction among all stakeholders); training the public;[8] market management (based on market principles); management by dispute (a political framework to resolve conflicts among the stakeholders' interests); and rational (discussions among the general public and policy makers in decision-making processes). The respondents asked were to rank the style from 1 to 5, with 1 representing "very little" and 5 representing "very much."

The survey results showed that the highest mean values for the ranking were for the command style (3.65) and for the market style (3.43). The author of the report concludes that based on these rankings by decision makers, a combination of command and market management styles prevails in the Iranian national system of innovation. Furthermore, the report states that, "From one side, it is observed that most innovation policies of the country are dictated from top to bottom and without extensive engagement of the stakeholders in decision-making processes, and from another side, because of the vacuum created by a lack of proper implementation of the governmental policies and ineffectiveness of government-decision rules, the market pressure influences the demand and supply of innovation. However, because of extensive market failures in innovation decision making, the development of innovation has been less than successful in the country" (IRPHL 2009, 204). This implies a lack of cohesive management of the national system of innovation, which may be a contributing factor to the country's inadequate innovation.[9]

To better understand the rift between science and production in the NSI of Iran, we reviewed the historical records of S&T development in the West, Japan, and the former Soviet Union to see the pattern of transfer of scientific achievements by large integrated firms to industries in the early stages of development[10] in technologically advanced economies. In an interesting study of the history of science and technology, Inkster (1991) observed that, "When considering the eighteenth century, there is little insight to be gained from an artificial separation of interests in natural science and interests in technique—the same publications and the same congeries were involved." Furthermore, he goes on to write, "An artisan inventor of 1750 or 1790 who read a scientific journal also imbibed explicit knowledge of new techniques" (50–51). However, an inventor-entrepreneur model of innovation was replaced by organized industrial research and the establishment of R&D units in large integrated enterprises in the late nineteenth and the early twentieth century. The creation of industrial research laboratories, agricultural experiment stations, and modern research universities, which are considered sources of invention, did not take place until the second half of the nineteenth century. These institutions made it possible to transform intellectual and physical capital into new knowledge, new technology, and ultimately wealth (Mowery 1990).

In the United States, where antitrust laws were vehemently enforced during the first decade of the twentieth century, many large enterprises relied on R&D for securing patents, which they could use to sustain or develop a dominant market position (Mowery 1990). Moreover, in addition to creating process and product innovations, the R&D units had another important function of monitoring technological opportunities and threats.

In any events, according to Chandler (1984), investment in industrial research never preceded the establishment of mass production and mass distribution. The pattern of development of large integrated firms, as described by Chandler, started with achieving significant economies of scale and scope (only industries with capital-intensive technologies could do so), organizing professional teams to manage production, and developing national and often international sales networks, and finally ended in the forming of a business organization that could coordinate the activities of these units. Establishing industrial research came last. In Chandler's words: "Investment in production abroad followed, almost never preceded, the building of an overseas marketing network. So too in the technologically advanced industries, the investment in research and development followed the creation of a marketing network. In these firms, this linkage between trained sales engineers, production engineers, product designers, and the research laboratory became a major impetus to continuing innovation in the industries in which they operated" (Chandler 1984, 492).

Based on these observations, we conclude that in-house R&D units and the invention of large integrated enterprises in the West were mostly endogenous to the working of managerial capitalism, and that the state played a very minor role. However, developments in Japan were somewhat different.

In the case of Japanese industrial development, both the state and the private sector played important roles. The Japanese government laid the foundation of industrialization through the construction of infrastructure, albeit in the face of fierce opposition by the public and conservative samurai, by building the first railroad lines and the nationwide telegraph network, which were completed in the 1880s. Moreover, the main vehicle of technology transfer to the private sector was government privatization of certain state-owned enterprises at "bargain-basement prices" in 1881. In discussing Japanese technological transformation, Morris-Suzuki (1994, 86) states: "Technological modernization was possible in Japan not just because of wholesale borrowing from the west, but because Japanese entrepreneurs and artisans developed intermediate technologies which fitted the existing economic structure. These intermediate techniques, then, were not the products of government technological policy, but were rational responses to market conditions in Japan, spurred on by exposure to competition from imported manufactured goods."

Of course, the motivation of the Japanese government for selling these state-owned enterprises at distressed prices was to manage massive government budget deficits, even though privatization in the second half of the twentieth century turned out to be a method by which many countries stimulated innovation. However, in nineteenth-century Japan, as was the case later in other countries where privatization took place, the chief beneficiaries of privatization were the political-economic elites. In Japan, the most immediate beneficiaries of the government transfer of SOEs to private ownership were a handful of powerful, wealthy individuals who had an ideological affinity and political allegiance to the Meiji political establishment. An example of such beneficiaries is Iwasaki Yataro, who is the founder of the Japanese giant multinational corporation Mitsubishi (Morris-Suzuki 1994).

As another example, one in which the circumstances are more similar to the Iranian case, we look at the problem of divergence between scientific output and industrial production in the former Soviet Union.

Searching for the cause of the rift between science and production, we find that one source of the problem may be the chasm that emerges between policy makers and science researchers on one hand, and industrial practitioners on the other. The chasm is a result of vast differences between the occupational environments of the individuals in the groups. The problem of the "bifurcation of science and industry" is a persistent one in all societies. In the industrial economies, the issue is resolved by bringing together science workers and industrial production professionals within large enterprises. Having the members of the groups working together does not eliminate all difficulties; however, it tends to reduce some of the problems (Amann and Cooper 1982).

Amann and Cooper (1982) in a massive historical study of Soviet innovation, found that many Soviet enterprises also faced the problem of keeping up the pace of innovation with enterprises in the West. In their examination of the problem, they found that in the former Soviet Union, "neither the structure of incentives nor the organizational framework encourages harmonization of scientific and industrial objectives. Soviet industrial enterprises tend to play only a minor role in research and development." The authors further go on to say: " There are notable exceptions" and that "by far the largest proportion of industrial R&D is carried out by specialist institutes which are separate in both an organizational and a geographical respect from industrial enterprises, their ultimate customers" (Amann and Cooper 1982, 15).

As an example of the geographic and organizational separation of Soviet industries, Amman and Cooper cite the example of the AKESR range of industrial control instruments, stating that, "The initial research was done in Moscow; the experimental plant was set up in Smolensk, 380 kilometers from Moscow and the final manufacturing plant was located in the Western Ukraine, 2200 kilometers from Kazan" (Amman and Cooper 1982, 16).

Based on our examination of national systems of innovation, we see that many Iranian university research groups and research laboratories exist in geographical and organizational isolation from state-owned enterprises. Most research centers are located in Tehran or other large cities throughout the country, and there are no organizational ties between the R&D organizations and the production enterprises.

Based on this brief historical review of the emergence and role of industrial research in the West and Japan, as well as the problem of the rift between science and production in the former Soviet Union, we conclude that one cannot expect a smooth transfer of knowledge created at government-run laboratories and university research centers to public and private enterprises. Given the similarities between the Iranian and the Soviet cases, a situation where the major industrial enterprises were state owned, and where the source of scientific development also came from the state, one should look into the Soviet experiences in this context and with the aim of learning valuable lessons. However, that is a subject that is outside the scope of this chapter.

Summary and Conclusion

In this chapter, we reviewed and compared the institutions, inputs, and outcomes of the national systems of innovation in Iran. We used a functional approach to the study of innovation systems. We found that the Iranian system of innovation faces the challenge of converting knowledge to wealth at the industry level. Moreover, the Western economic sanctions against Iran have created a bittersweet condition for the country. On one hand, the sanctions have reduced, if not completely eliminated, the transfer of technology from abroad. On the other, they have stimulated efforts in the country to rely on its human resources and domestic capabilities to develop S&T independent of imported technology. The indigenous technological achievements of the country will be discussed in the remaining chapters in this volume. These chapters indicate that the technological "shock" of sanctions to the country, instead of retarding its growth, has worked as a powerful force to advance technological development by leaps and bounds.

As Hekkert and colleagues (2007) suggest, job creation is the heart of national systems of innovation. Without it, the chain of R&D-innovation-value creation remains disjointed. Iran's NSI is no exception, and it suffers from the challenges of using innovation for job creation. The Iranian leaders have become aware of the problem, and have recently have been putting increased emphasis on the commercialization of innovations.

Facing the present economic sanctions and military threat from the West, if Iran is to achieve technological self-sufficiency, she must follow the Japanese path of industrialization, which is industrialization from above.

Notes

1. Roughly speaking, PPP refers to the value of foreign goods in terms of home-country currency units. For example, if it takes $10 to buy a basket of goods in the United States, and 1000 rials to buy the same basket of goods in Iran, then the PPP exchange rate between dollar and rial is $1 = 100 rials. In calculating the PPP exchange rate between two currencies, both the prevailing prices in the countries and the spot exchange rate between the two currencies are used.
2. For the Iranian government's policies in promoting entrepreneurship in nanotechnology in the country, see section 2.6.1 of the chapter on nanotechnology in this volume.
3. GEM defines TEA as the percentage of the population between 18 and 64, who are engaged in start-up business activities up to 3.5 years of business operations. For details and definitions, see page 24 of the Report.
4. For a history of Jondi Shapour University, see chapter 1 and the discussion of educational policies and developments in chapter 2 of this book.

5. In this study, we rely heavily on this excellent study of education, science, and technology policies and development in Iran, which is in the Farsi language.
6. Institute for Scientific Information (ISI), now a division of Thomson Reuters, publishes the impact factor or the number of citations for articles that are published in science and social science journals.
7. The data are based on personal observation of one of the authors as a consultant to the Ministry of Science, Research and Technology.
8. This view believes that conflict in decision making among the stakeholders is a reflection of the inadequacy of knowledge on the part of the general public. This view suggests that the general public should be educated about the nature of the science relating to the dispute.
9. On government and market failures, see the introductory chapter of this book.
10. According to Chandler (1984), the integrated industrial eneterprises as large, complex business entities emerged in the 1850s and the 1860s in Europe and America, and later in Japan. These are multifunctional enterprises that first integrated forward by investing in marketing and distribution, and then integrated backward by acquiring suppliers and control of raw materials. These integrated firms are manged by professional management teams.

References

Amann, R., and J. Cooper, eds. 1982. *Industrial Innovation in the Soviet Union.* New Haven, CT: Yale University Press.

Bascavusoglu-Moreau, E. 2010. "Entrepreneurship and the National System of Innovation What is Missing in Turkey?" United Nation University Working Paper Series #2010–030. World Institute for Development Economic Research (UNU-WIDER). Accessed December 29, 2011. https://www.google.com/#hl=en&output=search&sclient=psy-ab&q=Bascavusoglu-Moreau+United+Nation+University&oq=Bascavusoglu-Moreau&gs_l=hp.1.0.35i39j0i10i30j0i5.2783.2783.0.8539.1.1.0.0.0.0.254.254.2–1.1.0.les%3B..0.0…1c.__dRl3enarg&pbx=1&bav=on.2,or.r_gc.r_pw.r_qf.&fp=4e9e681d226667eb&biw=1280&bih=849

Chandler Jr., A. D. 1984. "The Emergence of Managerial Capitalism." *Business History Review* 58: 473–503.

Coghlan, A. 2011. "Iran Is Top of the World in Science Growth." New Scientist. March 28. http://www.newscientist.com/article/dn20291-iran-is-top-of-the-world-in-science-growth.html.

"Details of Scientific Articles Production in Iran" (in Farsi). 2011. Hamshahrionline. Accessed December 10, 2011. http://www.hamshahrionline.ir/news-156414.aspx.

The Expediency Discernment Council of the System. 2003. "Vision 1404" (in Farsi). Maslahat Nezam's website. Accessed September 25, 2011. http://www.maslahat.ir/DocLib2/Approved%20Policies/Offered%20General%20Policies/policy%2006–07–1382%20Iran%20Vision%201404.aspx.

Freeman, C. 1987. *Technology Policy and Economic Performance: Lessons from Japan.* London: Pinter Publishers.
Ghazinoory, S. M. Abdi, and K. Bagheri. 2010. "Promoting Nanotechnology Patenting: A New Experience in National Innovation System of Iran." *Journal of Intellectual Property Rights* 1 (6): 464–473.
Global Entrepreneurship Monitor (GEM). 2010. "2010 Global Report." Global Entrepreneurship Monitor. Accessed November 5, 2011. http://entreprenorskaps forum.se/wp-content/uploads/2011/02/GEM-2010-Global-Report.pdf.
Hekkert, M. P., R. A. A. Suurs, S. O. Negro, S. Kuhlmann, and R. E. H. M. Smits. 2007. "Functions of Innovation Systems: A New Approach for Analyzing Technological Change." *Technological Forecasting & Social Change* 74: 413–432.
Hekkert, P., and Simona O. Negro. 2009. "Functions of Innovation Systems as a Framework to Understand Sustainable Technological Change: Empirical Evidence for Earlier Claims." *Technological Forecasting & Social Change* 76: 584–594.
Inkster, I. 1991. *Science and Technology in History: An Approach to Industrial Development.* London: Macmillan.
Institute of Research and Planning for Higher Learning (IRPHL). 2009. "Final Report: Study Organs and Management of National System of Innovation" (in Farsi). Tehran: Ministry of Science, Research, and Technology.
———. 2010. "Report of Higher Learning, Research, and Technology of the Nation, Year 1387" (in Farsi). Tehran: Ministry of Science, Research, and Technology.
Iranian Research Organization for Science and Technology (IROST). 2011. Accessed on 21 October 2011. IROST.org. http://library.irost.org/persian/links/viewAll.asp?key=Institute)
Johnson A., and S. Jacobsson. 2000. "Inducement and Blocking Mechanisms in the Development of a New Industry: The Case of Renewable Energy Technology in Sweden." In *Technology and the Market: Demand, Users and Innovation,* edited by R. Coombs et al. 89–111. Cheltenham, UK: Edward Elgar.
Kelley, J. D., N. Bosma, and J. E. Amors. 2011. "Global Entrepreneurship Monitor 2010 Annual Report." Global Entrepreneurship Research Association. Accessed October 12, 2011. http://www.gemconsortium.org/download/1313156780237/GEM%20GLOBAL%20REPORT%202010rev.pdf.
Khamenei, Ali. 2005. "Address to University Professors and Elite Academics," October 13, 2005. Leader's Address to University Professors and Elite Academics. October 13, 2005, Office of the Supreme Leader, (access date, 21 October 2011) http://www.leader.ir/langs/en/index.php?p=contentShow&id=3472.
Malerba, F. 2002. "Sectoral Systems of Innovation and Production." *Research Policy* 31, no. 2: 247–264.
Mehr News. 2011. Report on Numbers of Students in the New Academic Year. Accessed December 10, 2011. http://www.mehrnews.com/fa/newsdetail.aspx?NewsID=1414082.
Morris-Auzuki, T. 1994. *The Technological Transformation of Japan: From the Seventeenth to the Twenty-First Century.* Cambridge: Cambridge University Press.

Mowery, D. C. 1990. "The Development of Industrial Research in U.S. Manufacturing." *American Economic Review* 80: 345–347.

Organization for Economic Cooperation and Development (OECD). 2005. "The Measurement of Scientific and Technological Activities: Guidelines for Collecting and Interpreting Innovation Data." In *Oslo Manual*, 3rd ed. prepared by the Working Party of National Experts on Scientific and Technology Indicators, para. 146. and para. 149. Paris: OECD.

Radosevic, S. 2005. "National Systems of Innovation and Entrepreneurship: In Search for a Missing Link," Project no. 506022, Centre for the Study of Economic and social Change in Europe, School of Slavonic and East European Studies, University College London (SSEES, UCL). Working Paper, number 73. CESPRI, Final Paper. Accessed December 6, 2011. http://www.ssees.ucl.ac.uk/csesce.htm.

Romijn, H., and M. Albaladejo. 2002. "Determinants of Innovation Capability in Small Electronics and Software Firms in Southeast England." *Research Policy* 31: 1053–1067.

Schwab, Klaus, ed. 2010. *The Global Competitiveness Report 2010–2011*. Geneva: World Economic Forum.

Shane, Scott. 2003. *A General Theory of Entrepreneurship: The Individual— Opportunity Nexus*. Cheltenham, UK: Edward Elgar, 2003.

Solow, R. 1956. "A Contribution to the Theory of Economic Growth." *Quarterly Journal of Economics* 70: 65–94.

Soofi, A., and S. Ghazinoory. 2011. "The Network of the Iranian Techno-Economic System." *Technological Forecasting & Social Change* 78: 591–609.

The Supreme Council of the Cultural Revolution. 2010. *The General National Scientific Plan* (in Farsi). Tehran.

Tabnak. 2011. "The Supreme Leader's Speech to a Gathering of Young Talented Individuals." October 4. Accessed October 10, 2011. http://www.tabnak.ir/fa/news/194891/.

United Nations Conference on Trade and Development (UNCTAD). 2005. *Science, Technology and Innovation Policy Review*. The Islamic Republic of Iran.

United Nations Development Programme (UNDP). 2011. International Human Development Indicators. Human Development Report 2011. Accessed August 9, 2011. http://hdr.undp.org.

Wikipedia. 2011. "List of Countries by GDP (PPP) Per Capita." Last modified August 21, 2012. http://en.wikipedia.org/wiki/List_of_countries_by_GDP_(PPP)_per_capita.

CHAPTER 5

Information and Communication Technology: Between a Rock and a Hard Place of Domestic and International Pressures

Sepehr Ghazinoory and Reza Jamali

Introduction

Applications of Information and Communication Technology (ICT) refer to the uses of computers, the Internet, traditional and mobile phones, as well as television and radio in activities such as agricultural production, education, healthcare, transportation, manufacturing and construction, financial services, customer and other business services, entertainment, and government services. Clearly, as evidenced from causal observations over the last three decades, ICT has profound impacts on the socioeconomic development of countries regardless of the degree of their industrialization.

ICT is a general-purpose technology (GPT) that has the potential for the fundamental restructuring of an economy. GPT, unlike other technical changes, is not incremental, and the successful adoption of the technology brings about a radical transformation of technological development, thus enhancing innovations in an economy in general. In short, GPT promotes innovations in all industries in an economy.

The ultimate goal of investment in ICT is the informatization of a society. Informatization is not a goal in itself, but is intended to achieve critical socioeconomic development goals.

The informatization process involves significant investment in economic and social infrastructure that is intended for the use of ICT by government, industry, and the general public as the stakeholders in the informatization processes.

ICT as the preliminary step in the informatization of a society impacts economies through a number of channels. First, investment in ICT contributes to capital deepening and raises labor productivity. Second, investments in and technological progress in the ICT-producing sector of the economy enhance efficiency of labor and capital; hence, they promote total factor productivity growth. Third, widespread diffusion of ICT in the economy raises total productivity growth of all sectors of the economy through the creation of intrafirm and interfirm networks, lowering transaction costs and creating conditions suitable for more process and product innovations (see, for example, Pilat 2004).

In short, outside of the output (employment) effect of investment in ICT as capital goods (the supplier effect), the uses of ICT could bring about process and product innovations and increase efficiency in production (the user effect).

Even though earlier studies showed a negative effect of ICT on productivity growth in developed economies (Pilat 2004), more recent studies show that the long-term growth contribution of ICT was significantly positive in the industries, both in the United States and the European Union (EU) over the entire period from 1980 to 2000 (Dimelis and Papaioannou 2011). These authors find that the productivity effects of ICT were mainly present in the industries, which were either ICT producers or heavy ICT users during the two decades of the study.

Studies dealing with the effect of ITC on the economic growth of Iran include Kamijani and Mahmood-Zadeh (2008), and Mahmood-Zadeh (2009). Kamijani and Mahmood-Zadeh (2008), using the annual data from 1959 to 2003, conclude that ICT contributed 7 percent per year to the economic growth of the country between 1994 and 2003. Mahmood-Zadeh (2009) shows that ICT makes an important contribution to the productivity of the factors of production in the country also.

Based on these earlier studies, one could infer that ICT would have a positive economic impact on the Iranian economy. Accordingly, we take a case-study approach in examining ICT policies, ICT development, and their effect in Iran. Specifically, after presenting a short history of ICT in Iran, we plan to examine the development of the ICT sector in Iran by critically reviewing the following categories:

- e-strategies and policies
- development and production
- investment and access

- diffusion and use
- performance
- evaluation

A Short History of Information and Communications Technology in Iran

Telegraph and Telephone

The first telegraph line in Iran was installed in 1857 when the city of Tehran was connected to the city of Chaman Solatneyah, in northwest Iran. Two years later, this line was extended to Zanjan, Tabriz, and Jolfa, and in turn was connected to Russia and, therefore, to the European telegraphic network (Center of Communications and Modern Technological Studies 2011).

The first telephone line was installed in Tehran in 1890, or 12 years after the invention of the telephone. However, widespread use of the telephone in the country only began in 1941. The telephone network that the Allied Powers had established in Iran during the occupation of the country during World War II was later purchased by the Iranian government after the Allied armies left Iran at the conclusion of the war.

The more recent development in the telecommunications industry in Iran started with the establishment of the first factory for telecommunications equipment in 1966. Then, in 1969, the first microwave communication system was put in place. Two years later, the first communication station was installed in Hamadan. Following this trend, the Iranian government, in cooperation with the government of Japan, set up the Iran Telecommunication Research Center in 1971. Moreover, that same year the Iran Telecommunication Company was founded, which currently is the main body that implements the telecommunication technology policies of the Islamic Republic (IR) of Iran. These organizations had the tasks of establishing the country's communications network as well as producing its communications equipment (Center for Communications and Modern Technological Studies 2011).

A period of low investment in the telecommunications infrastructure was the natural outcome of the revolutionary change, political instability, and invasion of Iran by Iraq. Nevertheless, during 1979–1984 efforts to expand and develop telecommunications included the setting up of more than 351,000 telephone lines, connecting 1,363 villages with the national telephone network, establishing long-distance services in 101 cities, and installing 4,083 local and long-distance public telephone booths throughout the country (Telecommunication Company of Iran 2012).

The development of communications technology in Iran received a boost when a factory for the production of optical fiber cables (OFC) was established in 1989 in Yazd, Iran. In that developmental spirit, the first long-distance coaxial cable link between Tehran and Isfahan, and then the first digital switching center and long-distance OFC link between Tehran and Karaj were established in 1988 and 1989, respectively. Two years later, Data Communication Company of Iran (DCCI) was founded for the purposes of planning, operating, and developing a data network, and the first Inmarsat Coast Earth Station began operations in the country. DCCI was able to set up the first phase of the Iran Public Data Network (PDN), with a X.25 protocol.

Continuing on the path of development, the Scientific and Applied Telecommunication College, which was closed after the start of the Islamic Cultural Revolution in the early 1980s, was reopened, and it merged with other universities in 1992. Additionally, Very Small Aperture Terminals (VSAT), which transmit and receive narrow and broad-band data to orbiting satellites, were set up by Telecommunication Company of Iran (TCI) in 1993. Moreover, digital switching centers in eight cities were set up, and the establishment of the Islamic Revolution Telecommunication Complex or Tehran Large-Capacity Tandem (LCT) took place in the same year also. In short, the TCI's crowning achievements in 1994 were the expansion of the Tehran local telephone network to 1,945,000,000 lines and the creation of the Tehran cellular network.

Given that most of the developments in the telecommunications infrastructure occurred in or around the vicinity of Tehran, the Islamic Consultative Assembly, the legislative branch of the Iranian government, aimed to develop that infrastructure throughout the provinces and ratified a law by which provincial telecommunications companies were established in 1995. By 1996, these companies began operations by expanding cellular networks in 35 cities to service one million mobile devices, by connecting 90 cities to the national data network, and by expanding the VSAT network by 70 percent.

Continuing on this telecommunications service expansion path, the Ministry of Communication and Information Technology authorized formation of the Mobile Communication Company of Iran (MCI) in 2004. As the name of the firm implies, this company was given the responsibility of creating the infrastructure network for the cellular telephone services in the country. By the middle of 2011, 134 cities, and fifty-two thousand kilometers of highways in Iran were covered by the cellular network. MCI has issued fifty-one million accounts with 1,148 operators, with a penetration ratio of 68 percent. Moreover, the company has established roaming connections with 267 operators in 110 countries, and the services it provides include voice mail, call waiting, caller ID, conference calls, fixed dialing number (FDN) services, mobile Internet connection, and all the other features of modern communications capabilities (Mobile Communication Company of Iran 2012).

It is noteworthy to point out that as the organization responsible for managing the telecommunications industry, in addition to developing the aforementioned infrastructure, TCI manufactures more than 80 percent of the needed telecommunications equipment inside Iran. Moreover, TCI has been very active in the export of technical and engineering services. The TCI completed the seven-hundred-kilometer Trans Asia Europe (TAE) OFC project in Turkmenistan, which involved planning, procurement of equipment, implementation of planned tasks, and training of personnel. The TAE has a capacity of 7,560 channels. Moreover, the backup route has 14 radio links, and contains 1,890 channels.

The Iranian telecommunications industry is among those of five countries that had 20 percent growth, and it has received the United Nations Educational, Scientific and Cultural Organization (UNESCO) award for its provision of telecommunications services to rural areas (TCI 2012).

Radio and Television

In 1924, the Ministry of War made preparations for use of radio in the country, and two years later, radio transmitters and receivers were imported to Iran. In 1934, the prime minister's cabinet approved the use of radios and passed a set of regulations according to which installing a radio antenna required a special permit from the Ministry of Post and Telegraph. The first radio station in Iran, with 8 hours of broadcasting, began operation in 1940. The broadcasts consisted of music, news, religious sermons, and cultural, geographic, and historical talk shows. (Wikipedia 2012a).

The first television station, named Iran Television, broadcast its programs in Iran in 1958. Initially, it broadcast daily for four hours from 6 PM to 10 PM. The Iran Television Company was privately owned and managed, and financed its expenditures through advertising revenues.

After one year of broadcasting, the television station increased its daily programming by one hour, opened another broadcasting station in Abadan, and installed a relaying station in Ahvaz.

The success of television broadcasting in Iran led the government to establish a national network of television broadcasting stations, and in 1964, the Plan and Budget Organization of Iran concluded an agreement with a French consulting group to plan and establish a network in the country. After adoption of the National Television of Iran proposal, the government started a new, small broadcasting station and began experimental programming two years later.

The name of Iranian television, National Television of Iran, was changed to Voice and Face (Seda va Sima) of the IR of Iran after the Revolution, and began broadcasting with two networks. During the first few years of operations in postrevolutionary Iran, the network broadcasting consisted of political documentaries and news only.

The government funds the activities of Seda va Sima through its annual budget. Other sources of revenues of the state-owned broadcasting enterprise come from advertising fees, fees charged to the consumers of electricity, and an earmarked sales tax on televisions that are purchased.

There were 14 country-wide broadcasting and 34 province-wide networks in Iran in 2012. Jaam-e Jam has networks, 1, 2, and 3, which are set up for Iranians abroad and cover Asia, Europe, Americas, and Australia. The Arabic-language networks Al-alam and Al-Kosar, as well as the English-language network Press T.V. are among the other networks that have global coverage. Moreover, networks 1 and 2 of Sahar, have daily programming covering 38 hours of broadcasting in Ordu, English, Bosnian, Turkish, French, and Kurdish. Additionally, Seda va Sima has 6 international radio broadcasting stations that broadcast in 25 foreign languages, including the station Voice of David, which broadcasts in Hebrew (Wikpedia 2012b).

Starting in 2009, digital radio and television broadcasts with DVB-T standard began in Tehran, and Iran has reached self-sufficiency in the building of strong radio transmitters.

Information Technology

The earliest use of software in Iran goes back to 1962; however, in the 1970s software found widespread use at universities and offices in the country. This period was marked by intensive activities in purchasing hardware and software, and in the hiring of experts in information technology. In the postrevolutionary period, and after the reopening of the universities in 1983, software development expanded, and many computers were acquired. One of the most important activities in software development during this time was the creation of word processing in Farsi. In the early 1980s, discussions relating to the export of software products took place, and by 1987 definite plans for the export of software were adopted. With the passage of the legislative mandate for development and with applications of information technology in 2003, a solid demand for software products was created, although most of the demand was for software for use by governmental agencies and ministries (Majlis Research Center 2010a).

After this brief historical review of the development of ICT in Iran, we turn to an examination of the ICT system in the country.

E-Strategies and Policies

Before the Iranian Revolution of 1979, Iran had invested heavily in ICT and its applications, which placed the country in a good position relative to other developing countries. However, after the Revolution, the government's attention to this technology diminished somewhat, and during the Iraq-Iran war, throughout most of the 1980s, efforts to further develop these technologies slowed to a certain degree. Nevertheless, in the last decade all sundry administrations have focused attention on and invested heavily in ICT. In this section, we focus on discussions of the e-strategy of the government.

The ICT Policy-Making Bodies

Technology strategy formulation in general, and ICT strategy formulation in particular, are the responsibilities of a few centers of power in Iran.

The Supreme Leader, Ayatollah Ali Khamenie, receives direct reports from the Expediency Council, the Supreme Cultural Revolution, and the Supreme Council of National Security. Each of the centers of power has its own ICT policies.[1]

The second center of decision making related to ICT is the Iranian Parliament, while the third is the cabinet of the President of the Republic, which has its own agencies such as the Supreme Council for Informatics, and the Supreme Council for Information Technology that deal with ICT policies. However, the implementation of the policies is the responsibility of the Ministry of Communication and Information

Technology. Of course, other government ministries get involved in policy making related to ICT matters that are internal to the ministry; however, they are not active in strategic decision making that relates to ICT in the country.

Numerous documents concerning the development of ICT in Iran have been developed over the last 20 years. We were able to assemble a comprehensive list of these documents, some of which are produced specifically for the development of ICT, and others of which are designed to deal primarily with other topics, even though they also make reference to ICT. In this chapter we examine the most recent, comprehensive General Plan for the formulation of e-strategy in Iran.

Overall, 25 documents relating to ICT were adopted in the country; however, the most important of these laws/policies is the Development Plan for Information Technology (TAKFA),[2] which was passed in 2003. It is estimated that some $500 million in funds were invested over several years to implement this law; nevertheless, some of the goals of the plan are yet to be met.

The E-Strategies under the Fifth Development Plan

E-strategy at the national level refers to a plan formulated by government leaders on how ICT should be used for the economic, social, and cultural development of the country. The document prioritizes the ICT projects and guides government investment of resources in the ICT projects, which are prioritized. Accordingly, the demand for ICT is a derived demand, that is, a demand for informatization of the society, which is based on the desire to achieve socioeconomic, cultural goals. In short, an e-strategy states what goals it is hoped will be achieved and how to achieve them.

Recognizing the importance of ICT in cultural, economic, social, political, and national security arenas, the Ministry of Communications and Information Technology (MCIT) concluded an agreement with the Management and Budget Organization to produce a General Plan for national information technology development in the country. The agreement involved planning for five main projects: Development of a General Plan for expanding information technology (IT); development of a plan defining the overall responsibilities of the MCIT in security issues; development of a database containing statistics about the condition of IT in the country; development of frameworks for the application of IT in Iran, and preparation of drafts for bills and regulations for legislative actions.

IT professionals, experts in IT strategy formulation, and individuals with expertise in management and professors[3] formed nine committees and after putting in twenty-one thousand hours of labor, they produced the first draft of the general strategy for the IT system in Iran. This document was adopted by Presidential Cabinet members on December 26, 2007.

The main strategy of the General Plan that was formulated by the committee is the "creation of a process with accelerating speed for closure of the gap between the states of IT development in Iran and that in the advanced industrial countries by using, developing, and deepening the national knowledge-based model; by acquiring a suitable share of IT-based goods and services from the global market; by providing for the needs of the dynamic, evolutionary society of the Islamic Iran" (Ministry of Communication and Information 2007, 8).

The Mission Statement

The mission statement of the General Plan reads: "Creation of access opportunities for all strata of the society to IT, and all-inclusive training of expert human resources for use in all aspects of life, and establishment of a creative, competitive environment for organizing a community of intelligent networks that can change the model and trends of resource-based development of the country to a knowledge-based development; with responsible citizenry in the acquisition of knowledge; and for elimination of the national digital divide with the global community" (MCIT 2007, 12).

The Committee developed strategies for seven focus areas of IT: "society and citizenry"; "development of high quality human resources"; "government and management methods for provision of needed, timely services to the public"; "R&D and technological innovation"; "R&D in IT"; "IT-based employment creation"; "creation of interactions among the national and supranational with the global networks of the Internet, companies, and markets" (MCIT 2007, 13).

In addition to the stated goals in the General Plan for ICT development, a number of specific IT functionalities at the national level were introduced in the Fifth Development Plan (2011–2016). These IT capabilities are listed below:

- establishment of direct electronic health files for the general public
- offering of the uniform health insurance services
- expansion of the national information network

- provision of electronic services by the government
- development of intelligent national identity cards
- setup of the national spatial data infrastructure
- establishment of a legal information database relating to real estate
- creation of a secure system for the legal transfer of real-estate titles
- provision of identity cards to legal entities such as corporations
- sharing of an organization's information
- establishment of infrastructure of the national scientific development network
- establishment of centers for issuance of electronic certifications
- defining of electronic certificates and their applications in electronic commerce
- establishing a centralized and interbanks operative electronic banking system
- expansion of electronic signing
- telecommuting

The plans for the development of ICT appear under different sections of the Plan and involve religious issues (promotion of Islamic-Iranian culture), science and technology, social issues (social security, social support and enabling services, health,) management (information technology, administrative issues), economics (reform of the tax system, capital market, improvement of the job market, money and banking, trade, religious giving or tithes), national defense, the legal system, and monitoring.

Development and Production

Review of the ICT industry in Iran will take place in three parts. First, we will examine the software industry; next we will review the communications industry; and finally, we will discuss the development of e-government. The discussion of e-government in Iran is particularly significant because of the prominent role of the Iranian government in the economy.

The Software Industry

The discussion of the development and export of applications software in Iran goes back to the 1970s, before the Revolution; however, the growth of this industry did not begin until adoption of the Development Plan for Information Technology (TAKFA) that took place in 2003. The development of the software industry rested entirely on governmental agencies and the private domestic market, since the Farsi language outside of Iran is spoken only in Afghanistan.

As another indicator of performance of the ICT industry in Iran, we cite the total exports of $124.5 million of software between 2001 and 2009, mostly to the United Arab Emirates and Canada. The exported software is mostly used by Iranian expatriates who live in this Persian Gulf country and North American country.

The domain of the activities of the Iranian software industry may be defined as consisting of the following software categories (Majlis Research Center 2010a):

Management and Commerce: Some activities in this field have taken place; however, no major innovations have occurred in this area. Most of software in this field is imported, and has been modified for use in the Farsi language.

Accounting and organizational resource planning: Extensive activities are taking place by a variety of software developers in these areas. Of course, in the case of ERP (Enterprise Resource Planning), much of the development is based on the software produced by large software companies such as Oracle and SAP.

Games and Entertainment: Most software for games is imported. However, due to the government's encouragement in recent years, many new software companies have emerged, which are totally engaged in the development of software for games that do not display violence and sexual conduct and are in harmony with the Iranian culture and value system. It should be pointed out that the motors of many of these domestically produced games are imported.

Software with art content: Software for web design is produced in Iran; however, software for modeling, graphics, and imaging are not developed domestically.

Content management: Much software for processing content, managing of documents, and editing files is produced domestically. However, many of these are software programs of foreign origin and have been modified for use by Farsi-speaking users.

Data Management: Software for classification, clustering, and data mining is not developed in Iran, and this class of software is imported.

Developmental: Software for configuration management, program testing, and compiling is not produced in Iran and is imported.

Education and Referencing: Much educational software is produced domestically, some of which are multi-media and interactive.

Software for specialized industries: There exist a few specialized domestically developed software programs such as for equipment management, industrial control, flight control, and medical control. Most of the software is acquired by importing the hardware in which the programs are used.

Network application: Network browsers and search engines are not produced in Iran.

Network management: Most of the software for network monitoring, system upgrades, and management of optical networks is not produced in Iran, and the needed software is imported.

Networking: Most of the access equipment, call centers, telephones, local area networks, routers, and switching are not produced domestically and are imported.

Operating environment: With the exception of the project for development of the Linux operating system in Farsi, no other attempts to develop other operating systems have taken place. This might be due to the superiority of Linux as an open-access operating system, compared to Microsoft Windows as a proprietary system.

Security and Protection: With the exception of software for server protection, software for secure financial and information transactions and antivirus and antihacking software are imported.

Device driver: There exists no domestically developed driver software for graphics cards, data conversion, data compression, and hardware drivers.

Data converter: Several software packages for electronic mail have been developed in Iran.

In summary, the discussion above clearly point to less than optimal performance of the software industry in Iran, given the human resource base of the country. In the evaluation section below, we will discuss some of the reasons for this subpar performance. Nevertheless, we should point out that the industry is currently going through a major transformation, and pivotal improvements are expected. This is due to the security concerns of the country and because software development for use by government agencies and ministries is rapidly developing.

Communications Equipment and the Computer Hardware Industry

We will examine the state of the communications equipment industry in Iran in this section. It is important to note that this industry produced many items such as telephones, broadcasting equipment, telecommunications equipment, computers, networks, and related equipment. Given the strategic importance of the communications industry, with an annual total global output of $200 billion, the governments in many countries have subsidized the industry, which has led to competitive advantages for many firms in these countries. However, the Iranian communications equipment industry, which consists of 80 public and private firms, did not have the same robust growth experience, and many of the firms in the industry are presently on the verge of bankruptcy.

Several issues have contributed to this state of affairs.

1. Unlawful import of competing products
   In spite of the laws prohibiting the import of much of the equipment that domestic firms are capable of producing, the unlawful import of these products has created formidable challenges for firms in the industry. For example, the annual import of cell phones in Iran is $2 billion; however, only about 5 percent of the imports come through Iranian Customs, and the balance is through illegal trafficking.

   It is instructive to mention that Iranian imports of telecommunications equipment through the official channels are from ZTE and HUAWI, two Chinese firms, and from Nokia; Siemens and Erickson; however, the illegal imports are mostly from China (Majlis Research Center 2011).

2. Selective enforcement of the Maximum Use of Domestic Production Capabilities Law
   In 1997, the Iranian Parliament passed a law that required government procurements of domestic products and services if these goods and services are domestically produced. Moreover, the law required that all government contracts should be given to Iranian-foreign joint ventures, with the Iranians having a majority interest. In contrast to the other sectors of the economy, due to the privatization of the communications industry in Iran several years ago, this law is no longer applicable to this industry.

3. The absence of a plan for industrial growth
   No development planning and strategy for industrial growth of this sector were formulated.

4. Inadequate interactions between research universities and the industry
   In spite of demonstrable technical capabilities in this field, as evidenced by the number of scientific papers that are published in high-quality journals by Iranian researchers in this area, the ICT companies in Iran do not take advantage of this expertise.

The E-Government

Similar to the situations in many other developing countries, full implementation of e-government in Iran faces a number of constraints, including first, the inability of many local and provisional governments to provide affordable, convenient information access points; second, low usage of e-government services, which is due to a lack of public awareness or the inability to use e-government services; and third, limited provision of e-government services.

Clearly, full, efficient development of e-government requires the coordination of supply and demand such services (Qiang 2007).

In general, e-government develops in 3 stages:

1. Creation of ICT infrastructure and incorporation of ICT applications into internal government processes (automation/management Information System (OA/MIS) stage)
2. Purposeful use of such applications in improving administrative functions (administrative and management stage)
3. Provision of e-government services to the general public (public services stage)

In a country like Iran, where the government plays a dominant role in the economy, the provision of e-government services is considered an important infrastructure, yet this infrastructure has evolved rather slowly. According to a United Nations' report in 2010 (United Nations 2010), in the Middle East and Central Asia region, Iran ranked very low in providing electronic government services: Only Iraq, Tajikistan, Turkmenistan, and Afghanistan were ranked lower than Iran. Nevertheless, the same report shows that Iran's global ranking in this field improved from a rank of 108 in 2008 to 102 in 2010, and there are signs that e-government services are accelerating.

More importantly, the index of e-government development for Iran, which was 0.38 in 2005, had a marginal gain, increasing to 0.42 in 2010. Following the same sluggish pattern of growth, three subindexes of the e-government index, that is, the index for human capital, online services, and infrastructure, changed from 0.74, 0.30, and 0.11 in 2005 to 0.79, 0.27, and 0.22, respectively, in 2010. The figures show a rapid growth of infrastructure development, and indicate promising robust growth in the provision of e-government services in the near future. (Majlis Research Center 2010b).

One of the issues that has caused the slow growth in providing e-government is inadequate cooperation among different government

Table 5.1 Trends of e-government in Government Agencies

	2004	2005	2006	2007	2008	2009
Government Agencies with a public integrated mechanized system	5	166	27	38	49	60
Government agencies with a private integrated mechanized system	1	5	9	13	17	20
Government agencies with a web presence	20	36	52	68	84	100
Government agencies with an interactive Internet system	0	5	9	13	17	20
Government agencies with e-services	0	3	6	9	12	15

Source: Majlis Research Center 2010b.

agencies and bureaucracies: They tend to show resistance in sharing information, and each agency considers the information it has as proprietary.

A survey study of government agencies and their electronic services resulted in interesting findings. The survey found a rapid rise in the indicators that appear in Table 5.1.

Investment and Access

The Iranian government's investment in ICT between 2002 and 2011 was more than $4.3 billion. As a reference point, the total fixed capital formation between 2000 and 2008 was 1,134,455 billion rials, where this sum at an exchange rate of $1 = 10,000 rials amounts to $113.445 billion (Bank Markazi Iran 2012). This sum reflects an annual growth rate of 0.5 percent of the government's annual budget. Moreover, the annual current expenditure on ICT constituted approximately 0.2 percent of the government budget, and approximately 0.33 percent of the government investment fund that was allocated for capital outlays in ICT.

Most of the government expenditures since 2006 have been directed toward infrastructure construction rather than spending for the supply of services: a ratio of 6:1 in favor of hardware acquisition. Furthermore, between 2002 and 2010, the per capita budget allocation for IT rose from $1.26 in 2002 to $10.51 in 2010, which indicates an eight-fold increase. However, in comparison, IT per capita spending in some of the countries in the region amounts to $1,000 per head, and in some advanced countries amounts to $2,000 per capita.

Additionally, over the last decade the share of IT of the per capita GDP fluctuated between 0.07 percent to 0.17 percent, which shows the relatively small contribution of ICT to the aggregate national output.

Comparatively speaking, the contribution of the only open-source software activities to the GDP of the European Union is 4 percent.

In line with the inadequate investment in ICT in Iran, which does not permit the system to enjoy economies of scale, the fixed cost of the infrastructure as well as the variable cost of supplying ICT services are substantially higher than those that prevail in many developed countries. For example, the cost of the subscription to one megabyte per second of Internet service in Iran is equal to the average monthly earning of Iranian citizens, while the subscription costs for the same in Turkey and in Japan are 1/11 and 1/894 of the average incomes in those countries, respectively.

In spite of inadequate investment in the ICT sector, the government has nevertheless promoted the industry by supporting private-sector firms in the industry. The most important backing is provided by the Ministry of Industry, Mines and Commerce, and the Ministry of Defense for the production and export of software and hardware. Moreover, the Office of Technology and Innovation of the President provides additional supports. Furthermore, the establishment of the Office of Deputy President for Science and Technology has invested heavily in assistance to innovative firms in this sector of the economy. Additionally, the Iranian Judiciary, by registering firms and patenting innovations, and by arbitrating commercial disputes plays an important role in this industry; however, many judges are not well informed about technical ICT issues.

In addition to direct investment in ICT and expenditure in support of private firms, the government spends considerable resources in the creation of industrial parks and R&D development centers. Furthermore, considerable resources are allocated for the development of human resources in the ICT field. ICT professionals are trained in high school as well as at the postsecondary level. Annually, a large number of graduates complete college and university degrees in the field of ICT; however, a large percentage of these graduates either immigrate abroad or cannot find suitable employment at home in their field of study. Finally, much of R&D in ICT takes place at universities and highly specialized ICT research centers such as the Center for ICT Research, the Ministry of Communication and Information Technology, the Center for Development of Electronic Commerce, and the Ministry of Industry, Mines, and Commerce (ICT Development Council 2012).

Diffusion and Use

In this section we examine the extent of the diffusion and use of ICT in the Iranian economy.

A study by Tavakol and Ghazinouri (2011) shows that the automotive industry in Iran had the best use of ICT, especially in areas of infrastructure and expert human resource development among all industries in the country. In some instances, the differences in the use of ICT between this industry and many European automobile firms are minimal. The natural gas and petroleum industry had adequate investment in ICT also, and is further developed in terms of ICT than other industries. The banking sector has invested heavily in ICT, even though the use of e-banking is not widespread, primarily due to depositors' mistrust of the system. However, the other sectors of the economy, particularly the insurance, leather and leather products, and agricultural industries require much improvement in the application and use of ICT.

Applications of IT in the health sector, especially in private healthcare institutions, have made substantial gains in ICT; however, the country's most important IT project that relates to healthcare delivery is the Intelligent Health Card (IHC), which is currently being used on a trial basis in some provinces. The IHC contains personal information about the cardholder, his or her updated medical history, health insurance information, information about the required urgent care, a list of prescribed medicines, and a personal identification number. The attending physician can access the data contained in the IHC, and prescribe medicine for the patient electronically.

Needless to say, the full, effective utilization of the e-health system hinges upon its availability in all physicians' offices, clinics, and hospitals across the country. However, due to the early stage of the development of the system, IHCs are not yet usable in many healthcare delivery centers. Nevertheless, currently e-health for twenty million residents in the rural areas of the country is in use, and the remaining population will received the card gradually.

Performance

We measure the performance of ICT in Iran by highlighting the major achievements of the country in this field.

National Information Network

Issues related to security and morally/religiously objectionable content on the Internet forced the government to initiate the development of a fragmented Intranet, a network with high bandwidth and numerous domestic data centers, what is called the "Halal Internet" in Iran.

Through this network, services such as databases, audio-visual IP_TV, social networks, e-government, e-business, e-education, and other functionalities will be offered. This network began its pilot run in 2011, and is becoming fully accessible during 2012.

Electronic Warfare Capabilities

During the time of national defense in the 1980s, Iran was able to capture some Western manufactured weapons from Saddam Hussein's army, and became acquainted with the concept of electronic warfare. In that period, an electronic warfare room was opened at Amir Kabir University in Tehran, and R&D in this field was initiated and subsequently expanded.

The most important Iranian achievement in this field is the acquisition of technologies to build passive radars with a five-hundred-kilometer range and drones. Moreover, the national air-defense system consists of an integrated communications network, which benefits from ten different communications channels.

Network Security

There exist private companies in Iran that specialize in the field of network security, and they have created a filtering system called *Separ*, or shield. They supply products such as firewalls, filters, secure discs, authentication, and software that limit or block outgoing virtual private networks (VPNs). In general, a lack of trust in foreign-produced security software has led to major achievements in software development in the area of network security and coding, particularly in the defense industry in Iran.

Fuel E-Card

For some time in the past, the Iranian government, which subsidized fuels for the vehicles in the country, faced a growing economic problem in continuing the program. The problem arose from a rapid rise in the number of private vehicles in use, and the inefficient fuel consumption of these vehicles on the one hand and the illegal trafficking of subsidized fuel to neighboring countries with a high price differential on the other. To remedy this expensive problem, the Iranian Petroleum Company introduced e-cards for use by the public in 2007. According to this fuel delivery system, each vehicle owner receives an e-card for use at the service stations, and is allocated a certain amount of rationed, subsidized fuel per month. At the time of the writing of this chapter in May 2012, the cost of a liter of gasoline is 4,000 rials (about $0.32). Motorists are free to purchase any additional quantity of fuel at the higher price of 7,000 rials or about $0.56 per liter. The owners of luxury automobiles are not subsidized, and they must purchase fuel at the higher price of 7,000 rials per liter, but face no restrictions on the quantity of fuel they may buy. The cost of the purchased fuel is automatically withdrawn from the cardholder's bank account.

Given increases in the prices of fuels over the last several years, this program has been very successful by creating a sharp decrease in fuel consumption. The project's cost was recovered in two weeks time!

As a result of the government's rationing and removing the subsidies available to all motorists, the consumption of gasoline has dropped by 50 percent. According to the Secretary of Subsidy-Management Organization, the daily consumption of gasoline without the rationing and without higher prices would have been 120 million liters. Instead, the daily consumption after rationing and the lifting of the fuel subsidy has dropped to 59.3 million liters (Farzin 2012).

Video Games

Imported violent, sexually oriented video games, which are not harmonious with the Iranian culture, motivated the creation of video games domestically that are more in line with the Iranian culture and value system. Accordingly, with the establishment of the Foundation for National Video Games (www.ircg.ir) in 2006, the country took the first few steps in the development of the software in this field. This foundation initially confined its activities to classifying imported games; however, recently it has developed video games that are based on the Iranian culture and Persian mythical heroes. These games have enjoyed high public demand. Of course, the illegal copying of DVDs and the incongruity between Farsi and the language used in these games and other prevailing languages around the world have created some constraints for global marketing of the products, hence restraining further development in this field.

National Scientific Network

The National Scientific Network (NSN) is the network that connects the universities and research centers in the country, and it is part of the National Information Network. This network makes all information at all databases at the universities and research centers available to the users. The most important of the databases belongs to the Center for Information and Scientific Documents (www.Irandoc.ac.ir), which gathers all master's theses, doctoral dissertations, and research reports in the country and makes them available to the public. Moreover, the Center performs interlibrary loan services for all universities and research centers in the country.

Super Computer

In 2011, two supercomputers with the capacity of 40 to 85 trillion calculations per second, and a calculating memory capacity of 7,500 gigabytes were manufactured at two universities in Iran. The connectivity speed of the supercomputers is 40 gigabytes per second. The supercomputer at Amir Kabir University has the capability of connecting to the national grid of the supercomputers in the country, and is in full control of the hardware and software components of the network. The network is capable of solving a collection of scientific and engineering problems that require complex computations. For this reason, from the beginning of the design of the project, a team of experts has been working on developing the software required by universities, research institutions, and private companies, and has developed and installed more than 60 software applications for the grid.

Some of the applications of the supercomputers in Iran include the simulation of combustion; propulsion systems for airplanes and spacecraft; the simulation of jet engines; the improvement of combustion and the reduction of engine pollution; prediction of damage to an automobile body after accidents, and improvement in automobile safety; simulation of the delivery of medicine to diseased cells in the human body; discovery of the characteristics of advanced materials by advanced molecular computations; calculations of atmospheric conditions; weather forecasting; the forecasting of natural disasters; and tests for submarines and ships.

The operating system for these supercomputers is Linux, which has been modified for use in Iran. Many specialized software programs have been installed in the supercomputers. In fact, by time this book is completed, Iran will have acquired IT capabilities for the design of large land and marine structures, large bridges, and aircraft, and the development of complex models of nano- and molecular technologies for use in the natural gas and petroleum industry.

Table 5.2 Indicators of Communication and Information Technology Development in Iran

Indicator	Iran 2000	Iran 2004	Iran 2009	Upper-Income Group 2009	Middle East & North Africa Region 2009
Separate Communication Regulator	–	–	Yes	–	–
Status of main fixed-line telephone operator	Public	Mixed	Mixed	–	–
International long-distance service	Monopoly	Monopoly	Monopoly	–	–
Mobile telephone service	Monopoly		Competitive	–	–
Internet service	Monopoly		Competitive		
Telcom. Revenue as a percentage of the GDP	1.2	1.1	1.4	3.3	3.1
Mobile cellular & fixed-line subscriber per employee	221	304	913	576	880
Telcom. invest as percent of revenue	6.0	122.8	74.5	18.0	23.6
Telephone lines (per 100 people)	14.8	21.9	35.4	22.1	15.8
Cellular phone subscription (per 100 People)	1.5	5.1	72.1	100.6	67.4
Fixed Internet subscribers (per 100 people)	0.4	–	–	19.4	2.0
Personal computers (per 100 people)	6.3	7.5	10.6	11.2	5.7
Internet users (per 100 people)	1.0	6.5	38.3	34.6	21.5
Population covered by mobile cellular network (percent)	32	–	95	94	93
International Internet bandwidth (bits/second/person)	1	15	151	1,120	323
Resident fixed-line tariff (US$/month)	10.8	2.8	0.2	10	3
Cellular prepaid tariff (US$/month)	–	2.9	3.6	8.8	6.3
Fixed broadband Internet access tariff (US$/month)	–	5.9	30.5	18.8	22.7
E-government Web measure index (0<index<1)	–	0.16	0.26	0.35	0.22
Secure Internet servers (per 1 million people, December 2010)	0	0.2	0.7	32.2	2.4

Source: World Bank 2012.

With the development of these supercomputers, Iran has joined a handful of countries, including China, Canada, Germany, France, Japan, and the United States that have the technology to produce supercomputers. The supercomputer that was made at Amir Kabir University of Technology has been listed among 500 superior computers in the world, and is ranked 108 among all supercomputers in the world.

Other Achievements

In addition to the achievements listed above, we provide a summary of the ICT achievements and rankings for Iran for the upper-income group of countries, and for the countries in the Middle East and North Africa in Table 5.2. Whenever available, we present the data over the selected years in the first decade of the twenty-first century.

According to the data in Table 5.2, in most categories, excluding telecom revenue as a percentage of the GDP, Internet bandwidth, e-government, fixed broadband Internet tariffs, and secure Internet servers, Iran outranks both the upper-middle-income and the Middle Eastern and North African countries.

Evaluation

ICT is not a common technology in the sense that it differs from other technology because it constitutes the infrastructure for both socioeconomic development and for the development of many other technologies. Moreover, ICT closely interacts with knowledge and has a social context. Development of ICT is as much dependent on technical and hardware needs as it is dependent on cultural and software needs.

Iran, being an Islamic state, has its own peculiarities and priorities that are influenced by several factors. On the one hand, the Islamic seminaries, which train religious scholars, are pioneers in the use of IT in their research and scholarly activities and have established extensive portals for distribution of their scholarly contributions. On the other hand, the political establishment considers this technology as a nuisance, a tool that is used for the dissemination of Western culture, for the creation of political instability, and for espionage as well as subversion in Iran. This contradiction has manifested itself in different political, legal, and educational dimensions, and as a result, the IT in Iran has had an unbalanced development.

To resolve this contradiction, as was discussed above, the government is in the process of establishing an independent national network for the Internet, which is congruous with the Islamic moral values of the society and which establishes a mechanism to guarantee national security concerns.

Below, we present some of the major reasons for the unbalanced development in ICT in the country.

1. Extensive governmental bureaucracies have acted as impediments in implementing the adopted goals.
2. Lack of complete respect for the principles of intellectual property, which is a reflection of the cultural and religious values (Islamic juirisprudence recognizes common property doctrine in certain casesrather than recognizing a private-property doctrine in all domains) of the society, has lead to extensive copying of software.
3. The government has imposed restrictions on the use of the Internet by individuals and households by supplying a maximum 128 kilobytes of Internet access to homes, and by filtering the political content of the Internet.
4. The Western economic and technical sanctions against Iran have made access to the latest advanced technology very difficult and, in some cases, impossible.
5. Outward migration of many expert IT professionals on the one hand, and the desire of those professionals who remain to work on assembling imported hardware have compounded the problem on the other.
6. Multiple ICT policy-making centers have shown interest in different aspects of ICT, with some interested in technological aspects (Ministry of Communication and Information Technology), some in cultural aspects (the Supreme Council of Cultural Revolution), and some in legal and security aspects (National Internet Refining Council).
7. The strong presence of public and quasi-public enterprises in the communications and software industries has resulted in less than optimal participation of the private sector in various industries and has eliminated all competition and innovation in those industries.
8. There is a limited number of Farsi-speaking customers in the region and the world.
9. The inconsistency between the academic programs and the employment needs of industry has confounded the problems in the ICT sector of the country.
10. Lack of a procurement system for governmental purchases that could support domestic production did not help in the promotion of industry.

Perhaps the difficulties listed above are not insurmountable. In spite of many bureaucratic impediments, a suitable infrastructure for access to the Internet and e-government is being created. The required laws have been passed, and the required human resources have reached a critical mass. It appears that the collective effect of these forces has created a suitable path for the development of this industry in Iran.

Conclusions

In the final section of the chapter, we use political, economic, social, and technology (PEST) analysis in assessing the ICT conditions in Iran.

The Technical Factors

The technical aspect of ICT development in Iran has faced some problems. Many well-trained IT professionals who have been educated in Iran are employed at substantially lower compensation at home in comparison to their counterparts in the advanced industrial countries. One may cite the many Iranian-trained IT professionals who are employed at leading IT enterprises such as Google, as well as the superior performance of Iranian high school and college students in international robotic competitions as evidence for the high quality and technical expertise of Iranian IT professionals. Hence, the combination of high training and low compensation has led to lingering problem of brain drain in this filed.

Political factors

The most challenging problems for the full development of ICT fall under this category. The national security and moral concerns on the part of the Iranian government authorities regarding potential abuse of the Internet by foreign powers to wage a cultural war and create political instability in Iran have led the authorities to implement a severe filtering of the Internet. Moreover, the Western-imposed sanctions have placed severe limitations on foreign investment in this field and on the purchase of many IT capabilities from abroad in the short-run. Nevertheless, one should not ignore the fact that this may be a blessing for the further, indigenous development of technology in Iran over the long-run.

It appears that as a result of the government suspicion of ICT applications in Iran, this sector, contrary to the other technological sectors, does not receive wholehearted support from Iranian leaders. Accordingly, the ICT sector in Iran faces both international and domestic countervailing pressures for more rapid developmental processes. Perhaps for this reason, a myriad of political, religious, cultural, security, and military organizations are engaged in ICT decision making in the country.

Economic Factors

The most important economic issue affecting ICT development is the lack of competition in the industry. Government is the most important player in the industry in terms of investment, consumption, and imports. Naturally, the government's presence in the industry has a profound impact on the private, small firms. Moreover, due to the inadequate investment in the ICT infrastructure, the average cost of providing Internet services is very high, and as a result the cost to Internet users of subscribing to this service is comparatively prohibitive (see Table 5.1 above).

Social and Legal Factors

There are several issues related to ICT in these categories in Iran. First, the culture of software piracy and the violation of intellectual property rights are prevalent. This tendency has a historical root. Ghazinoory and colleagues (2011) show that before the Islamic Revolution in 1979, Ayatollah Khomeini did not recognize intellectual property rights, and it was only after the triumph of the Revolution that he changed his view on this issue and recognized them. Given the cultural value system, many users of the software developed in Iran violate intellectual property rights, and developers, in attempts to stop the piracy, lock the software, which often causes difficulties in purchasers/owners using the software at a later date.

Second, given the fact that the Farsi language is used only in Iran and a part of Afghanistan, for all practical purposes, the usage of software developed in the country is confined to the Iranian domestic market. Hence, economies of scale in software development cannot be achieved by relying on the domestic demand only.

In the final analysis, the most promising indicator for the development of ICT in Iran is the desire and pressure by different strata of the society for the industry to advance further. A large proportion of the Iranian population is young, and was born after the Revolution. These youths are highly educated and create strong demand for electronic and Internet services, and the government as the sole provider of these services must feel compelled to meet the high demand. Otherwise, it will create antagonism and social conflict in the society. Creation of the National Information Network is an attempt to address this contradiction. In any event, the issue does not end at this juncture: It is hoped that the social demand, coupled with development of the existing or emerging ICT technological capabilities, will usher in a major transformation in the development and applications of ICT in Iran.

Notes

1. In March 2012, a new organization named the Supreme Council of Virtual Space was formed. However, the main functions of this new organization are not known.
2. TAKFA is the acronym for the name of the plan in Farsi.
3. It is interesting to note that contrary to the Western practices of including individuals to represent the interests of business communities and labor on committees set up by the government to formulate public policies, this committee did not have such representation.

References

Bank Markazi Iran. 2012. *National Expenditure at Constant Prices.* Bank Markazi Iran (Central Bank of Iran). Accessed May 10, 2012. http://www.cbi.ir/.

Center of Communications and Modern Technological Studies. 2011. *Report of Survey of Communication Equipment Industry in Iran* (in Farsi). Tehran: Center of Communications and Modern Technological Studies.

Dimelis, S. P., and S. K. Papaioannou. 2011. "ICT Growth Effects at the Industry Level: A Comparison between the US and the EU." *Information Economics and Policy* 23 (1): 37–50.

Farzin, M. 2012. "Use of Gasoline before and after Removal of Fuel Subsidies." *Online News* (in Farsi). Accessed April 26, 2012. http://www.khabaronline.ir/detail/209437/.

Ghazinoory, S., S. Ghazinoori, and M. Azadegan-Mehr. 2011. "Iranian Academia: Evolution after Revolution and Plagiarism as a Disorder." *Science & Engineering Ethics* 17: 213–216.

ICT Development Council. 2012. Report of ICT Budget Analysis. ICT Development Council. Accessed August 23, 2012. http://ictc.isti.ir/index.php?option=com_content&view=article&id=356&Itemid=236.

Kamijani A., and M. Mahmoudzadeh. 2008. "The Role of Information and Communication Technology (ICT) on Economic Growth in Iran: Growth Accounting Approach." *Economic Research Review* 8 (2): 75–107.

Mahmoudzadeh, M. 2009. The Effect of Information Technology on Labor Productivity in the Iranian Manufacturing Industries: 2002–2007." *Quarterly Journal of New Economy* 5 (17 and 18): 1–22.

Majlis Research Center. 2010a. "Pathology of Software Industry" (in Farsi). Publication no. 10434; subject code: 280. Islamic Parliament of Iran, Tehran. Accessed August 23, 2012. http://en.parliran.ir/.

———. 2010b. "E-Government Assessment" (in Farsi). Publication no. 10283; subject code: 280. Islamic Parliament of Iran, Tehran. Accessed August 23, 2012. http://en.parliran.ir/.

———. 2011c. "Report of the Investigation of the Communications Equipment Industry in Iran" (in Farsi). Publication no. 10770; subject code: 280. Islamic Parliament of Iran, Tehran. Accessed August 23, 2012. http://en.parliran.ir/.

Ministry of Communication and Information. 2007. The General System of Information Technology in Iran (in Farsi). Ministry of Communication and Information, Tehran.

Mobile Communication Company of Iran. 2012. A Short History". Mobile Communication Company of Iran, Tehran Accessed April 16, 2012. http://www.mci.ir/web/guest/history.

Pilat, K. 2004. "The ICT Productivity Paradox: Insights from Micro Data." *OECD Economic Studies* 38. Accessed August 23, 2012. http://ictlogy.net/bibliography/reports/projects.php?idp=1849.

Qiang, C. Z. 2007. *China's Information Revolution: Managing the Economic and Social Transformation*. Washington DC: The World Bank.

Tavakol M., and R. Ghazinouri. 2011. "Diffusion and Obstacles to ICT Adopting in Iranian Industries; Case Study of Selected Sectors." *Journal of Science and Technology Policy* 3 (2): 31–48.

Telecommunication Company of Iran. 2012. "History of Communication in Iran." Telecommunication Company of Iran, Tehran. Accessed May 20, 2012. http://tci.ir/about/index.aspx?lang=En.

United Nations Public Adminstration Program. 2010. "E-Government Survey 2010." UNPan.org. Accessed May 11, 2012. http://www2.unpan.org/egovkb/global_reports/10report.htm.

Wikipedia. 2012a. "Radio in Iran." Accessed August 23, 2012. http://fa.wikipedia.org/wiki/.

———. 2012b. "Seda va Sema of the Islamic Republic of Iran." Accessed August 23, 2012. http://fa.wikipedia.org/wiki/.

World Bank. 2012. "Information and Communication Technology (ICT) At-a-Glance Tables." Worldbank.org. Accessed August 23, 2012. http://devdata.worldbank.org/ict/irn_ict.pdf.

CHAPTER 6

Nanotechnology: New Horizons, Approaches, and Challenges

Fatemeh Salehi Yazdi[1] *and Mohammad Ali Bahreini Zarj*

Introduction

In recent decades, the Iranian Supreme Leader and various presidents of the Islamic Republic (IR) of Iran have emphasized the development of science and technology within the country. As a part of supporting policies and laws for technology development, Iranian policy makers have stressed the importance of emerging technologies such as nanotechnology. The reason for emphasizing nanotechnology in particular is its wide applicability in various production processes and the societal implications of the technology in many domains including economics, health, and the environment (Ghazinoory and Ghazinouri 2009). In addition to the important multifaceted usefulness and societal implications of nanotechnology, its development in Iran warrants serious study for another reason. For the first time in Iranian history, and particularly after Iran's revolution in 1979, the very advocates and policy makers who were focused on the technological development of the country were able to plan how to enter into an emerging technology field before the country faced a wave of imported products produced with that new technology. Accordingly, reviewing these experiences should both provide useful lessons and models for the development of other technologies in Iran and benefit technology policy-making practices in other developing countries.

This chapter is structured as follows: The next section will define nanotechnology and give a brief history of its development and structure as an emerging technology in Iran, as well as note key policies and identify the main actors. The third section examines the major challenges in this growing field, including the internal and external challenges within the industry. Following this, the fourth section looks at the performance of the key national developmental organization involved, the Iran Nanotechnology Initiative Council (INIC), in terms of its human resources, business environment, and international interaction and contributions to the development of this technology. Having considered these areas, the chapter notes what policy-making lessons can be learned from the successes and failures of this model, and presents its recommendations for further improvement of technology policy decisions.

Structure of Nanotechnology Development in Iran

Definition and Importance of Nanotechnology

A nanometer is one billionth of a meter (10^{-9} m)—about one hundred thousand times smaller than the diameter of a human hair, a thousand times smaller than a red blood cell, or about half the size of the diameter of DNA. Nanotechnology involves research and technology development at the atomic, molecular, or macromolecular levels using a length scale of approximately 1 to 100 nanometers in any dimension. Furthermore, the technology involves the creation and use of structures, devices, and systems that, due to their small size, have the ability to control or manipulate matter on an atomic scale (US Environmental Protection Agency 2007). Lux Research Inc., an independent research and advisory firm providing strategic advice and ongoing intelligence on emerging technologies, defines nanotechnology as "the purposeful engineering of matter at scales of less than 100 nanometers to achieve size-dependent properties and functions" (Lux Research Inc. 2006).

Nanotechnology presents opportunities to create new and better products. Nanotechnology revolutionizes manufacturing processes and has nearly boundless applications (Fulekar 2010). Products and applications of nanotechnology will have many implications and create tremendous changes in all aspects of human life. Purified water, clean air, and new medicines are three key examples of the benefits that applied nanotechnology can bring to society. These are important issues that many countries are facing, and the United Nations noted them in its Millennium Development Goals (UN Millennium Development Goals 2010). Governments and private entities everywhere are pouring billions of dollars into nanotechnology research and development each year (Roco et al., 2010). Similarly, Iran has been investing in this technology with the aims of creating wealth and improving people's quality of life.

Background of Nanotechnology Development in Iran

The development of nanotechnology in the country began in 2001 with technology-monitoring reports from some expatriate Iranian scientists in United States to the former Iranian president, Mohammad Khatami. At that time, Iran was seeking technology focus areas with high economic potential to build the country's industrial base. The issue was considered by the Technology Cooperation Office (TCO),[2] which is under direct supervision of the Office of the President, which is responsible for technology development in the country. TCO decided to pursue the development of nanotechnology for several reasons. First, it could have a significant impact on people's quality of life; second, there were large potential markets for products manufactured with the use of nanotechnology; and third, it had the goal of narrowing the technological gap between Iran and advanced industrial economies in the development and use of this technology. The third factor implies that Iran had a good chance to catch up with the leading countries in this technology in a short time span (Ghazinoory et al. 2010).

TCO created a committee to carry out studies related to nanotechnology development policies. After the initial studies and promotion of nanotechnology among specialists and experts, the committee recommended creation of a council responsible for the development of nanotechnology. In 2003, the INIC was established, and subsequently, the ten-year strategic plan for nanotechnology development, named "The Future Strategy," was approved by the Cabinet in July 2005 (INIC 2005). As a result, the INIC became the main national body responsible for implementing the plan (Ghazinoory et al. 2010).

The vision of the program includes using nanotechnology for creating wealth, improving the quality of life in Iran, and achieving a desirable position among the 15 countries that are most advanced in nanotechnology by 2015. During the first few years of implementing the plan, emphasis was mostly put on human resource development and infrastructure creation. Also, to provide researchers with suitable laboratory services and to strengthen the necessary research infrastructure, Iran Nanotechnology Laboratory Network (INLN) was established (http://www.nanolab.ir). The INIC started to support companies that were active in nanotechnology research and development in 2007. Most of these companies were university spin-offs or start-up entities that had concentrated on a specific field of nanotechnology development, and had launched their products in the market already (Ghazinoory and Farazkish 2010). The INIC has supported the establishment of some companies through venture capital firms as well.

After five years of program implementation, Iran now has significant achievements in this emerging technology. First, the country is ranked fourteenth in nanoscience development in the world based on the number of published articles that appear in the International Science Index (ISI), a service provided by Thomson Reuters Company (Maghrebi et al. 2011). Second, several patents have been registered in internationally recognized patent offices. Third, more than 140 companies and start-up firms are active in nanotechnology in Iran, and around 60 of them have launched nanoenhanced (or nanoembodied) products in the market as of 2009 (INIC Annual Report 2009).

Statement of National Policy on Development of the Technology

As stated earlier, the statement of national policy endorsing the development of nanotechnology is known as the Future Strategy document, and the INIC is responsible for implementing it. The Future Strategy also defines some nanotechnology development programs for eight ministries and governmental organizations. The INIC is in charge of coordinating the nanotechnology development activities of other governmental institutions, establishing nanotechnology development and use as a priority in these entities, and mobilizing their financial as well as human resources for the development of that technology.

The Future Strategy includes the objectives and the framework for implementation, funding, and evaluation of the plan. Its policies define which nanotechnologies match national priorities, articulate how to promote nanotechnology within Iranian culture, support academic research and provide pathways to develop the science and technology infrastructure.

The nanotechnology development strategy included planned revisions of the strategy according to changing external conditions. Therefore, it has been revised twice based on the new conditions, and its current version puts more emphasis on the requirements of advancing nanotechnology in the international arena and thus addresses areas such as commercialization, industry involvement, market development,

Table 6.1 National Priorities in Nanotechnology

Large-scale Priorities	Examples of Subdivisions
Energy	• Energy saving • Solar cells • Fuel conversion • Chemical cells quality improvement
Health	• Novel Drugs • Diagnostic kits
Environment and Water	• Water purification • Desalination • Wastewater treatment
Materials	• Nanocomposites
Construction	• Anticorrosion process • Strengthening of buildings and infrastructures

(table 6.1) patent registration, standardization and safety, production according to international standards, regional as well as international cooperation, and participation in international events.

In 2007, when the second phase of the Future Strategy document was compiled, the top priority issues were determined by taking into account the following general considerations:

- large-scale scientific and technological priorities of the country;
- needs, challenges, and advantages of the country;
- prospective opportunities in nanotechnology.

Based on these considerations, the following functional fields were chosen as nanotechnology priorities:

The Annual Budget of the Organization

Since 2004, the INIC has had a budget line in the annual budget law of the country, with the average annual budget of around $16 million as of 2009. The distribution of the budget for the year 2009 shows that more than half of the budget has been allocated to "Support for Study and Research" in order to encourage researchers at Iranian universities and research centers to focus on nanotechnology and publish their scientific achievements internationally. Less than 10 percent of the budget each year has been allocated to the "production and market" program.

Actors in the Nanotechnology Field in Iran

Major players in the nanotechnology field in Iran can be divided into three groups: companies active in the nanotechnology field, specialists, and the government. This section discusses the first two of these: nanotechnology companies and experts in the field. The role of government is explained in the section below titled Actors in Policy Formation for the Nanotechnology Field.

Nanotech Companies

To describe business actors in the field of nanotechnology in Iran, we use the nanotechnology value chain (Lux Research Inc. 2006; Bahreini et al. 2012). This value-chain structure recognizes the role that nanotechnology applications play from raw material through the final goods. Nanotechnology applications are categorized into one of four categories, three of which form a linear value chain and a fourth that spans that value chain.

The categories consist of:

- **Nanomaterials.** Nanomaterials are purposefully engineered structures of matter, with at least one dimension of less than 100 nm, exhibiting size-dependent properties that have been minimally processed. Nanomaterials exhibit unique properties attributable to their size. These materials are expensive to produce, and are not useful in their current state. Instead, they become valuable when they are incorporated into products further down the value chain to deliver desired properties.
- **Nanointermediates.** Nanointermediates are intermediate products. They neither appear in the first nor in the last step in the value chain. They either incorporate nanomaterials or have been constructed from other materials to have nanoscale features.
- **Nanoenabled products.** Nanoenabled products are finished goods at the end of the value chain that incorporate nanomaterials or nanointermediates.
- **Nanotools.** The previous three value-chain stages flow into one another—nanomaterials are used in nanointermediates, which are incorporated into nanoenabled products. Researchers and manufacturers working at all three of those stages, however, make use of the fourth value chain category, nanotools, in their research and development (R&D) and production activities. Nanotools are the capital equipment and software used to visualize, manipulate, and model matter at the nanoscale.

Based on the value chain, major actors in these four categories in Iran are described below.

- **Actors in nanomaterials category:** Around 50 private start-up companies have been established in the field of nanomaterials, the majority of which are based in incubators. Most founders of these start-ups are university graduates, and a few are specialists in the industry. Products of these start-up firms include various nanomaterials such as nanosilver particles, carbon nanotube, nanozinc oxide, nanotitanium oxide, and nanodiamond particles.
- **Actors in nanointermediates category:** Five companies are active in this category. These companies are mostly working on the applications of nanosilver particles in fabrics and ceramics and nanocomposites.
- **Actors in nanoenabled products category:** There are eight companies producing nanoenabled products. Their products include antistain clothes, antibacterial, antiodor and antifungi air filters, silent water pipes, motor oil, low-emission glass for buildings, long-lasting traffic paints, insulator paints for building, antibacterial wound sprays, and organic fertilizers.
- **Actors in nanotools category:** There are eighteen equipment- and instrument-producing companies that are capable of manufacturing different nanotechnology equipment such as scanning tunneling microscopes (STMs) and atomic force microscopes (AFMs).

In each value chain, there should be a proper balance in the number and size of companies in different categories. Investigation into the activities of Iranian nanotech companies shows that most companies are active in the upstream part of the value chain, producing nanomaterials. This is due to the fact that most of these companies are established by university graduates[3] in this field. In comparison with nanomaterial producers, there are few companies that are active in the nanointermediates category. In other words, there have been few integrated activities aimed at finding formulations for applying nanomaterials in the industry. Therefore, there are quite a few opportunities for entrepreneurs to develop nanotechnology applications and respond to the demand for nanointermediate companies. Filling the gap of the nanointermediates category will, in turn, lead to the development of more nanoenabled products and with higher qualities.

In their interactions with nanotech companies, the authors have witnessed that those industrial companies that invested in making nanoenabled products enjoyed higher rates of return on their investment than those companies investing in the upstream part of the aforementioned value chain, namely nanomaterials. The profitability of launching nanoenabled products caused their producers to move toward investing in the upper parts of the value chain, that is, the nanointermediates. To guarantee the supply of nanomaterials for more profitable companies that produce nanoenabled products, and to increase the competitive advantage of the firms in the industry, integration (vertical mergers and acquisitions) of the companies producing nanomaterials and nanoenabled products may be a policy of choice, which could enable the firms in the industry to compete globally.

Nanotechnology Specialists
Nanotechnology experts play a prominent role in the development of nanotechnology. Thus, the status of nanotechnology specialists in both technological fields and the business arena is explored here.

In 2001, when Iran began working on nanotechnology development, there were less than ten technical experts in this field in the country. To train such personnel, several master's and PhD programs were initiated at Iranian universities, and financial incentives were provided for researchers in nanotechnology. Currently, 14 Iranian universities have master's programs in the areas of nanomaterials, nanophotonics, nanoelectronics, nanochemistry, nanophysics, chemical engineering, nanopolymers, nanostructure filaments, nanobiotechnology, and nanomedicine. Also, 7 universities have doctoral programs in different specialties of nanotechnology.

The INIC has both horizontal and vertical policies to boost research in the field of nanotechnology. Horizontal policies include financial incentives for MSc and PhD research projects, scientific publications, and defraying the costs of obtaining patents in nanotechnology. This policy has targeted curiosity-oriented research that focuses on nanotechnology and its related disciplines. Vertical support policies include the support of research for selected issues of high priority for the country and the aim of having patents, products, or start-up companies as outputs. The outcomes of research-support policies will be discussed in detail below..

In spite of the fact that there is substantial number of technical specialists in nanotechnology in the country, there is a lack of business expertise required for nanotech business development. This is one of the major challenges of nanotechnology companies in Iran (Bahreini and Salehi 2010a). They face several problems in planning their commercialization activities, and often end up continuing on a trial-and-error basis. Although the Future Strategy document emphasizes the training of nanotech business experts, there were few instances of this in practice.

Actors in Policy Formulation for the Nanotechnology Field

As mentioned earlier in this chapter, based on the Future Strategy document, the INIC is the main body responsible for formulating policy regarding nanotechnology development in Iran. Those sections of the INIC that are in charge of policy formulation and implementation for the technological sector fall into two categories:

- Nanomaterial, nanointermediate, and nanoenabled products, which includes the Iran Nano Business Network (INBN), since its restructuring as the Technology-Market Services Corridor (TMSC); and
- Nanotools, which includes the INLN.

In the following sections, we discuss these categories in greater detail.

Technology-Market Services Corridor (TMSC)

To encourage the private sector to invest in nanotechnology and strengthen active nanotech companies, the INBN was established in 2007. The services provided by the INBN for nanotech companies include:

1. Cooperation with governmental or private funding institutions in securing loans for nanotech companies
2. Financing to support research and development in nanotech companies with the aim of expanding production and introducing new products to the market
3. Holding training courses and workshops to improve the managerial capabilities and business competencies of nanotech companies;
4. Participating in international exhibitions with the aim of introducing Iranian companies' capabilities to the visitors, establishing connections with nanotechnology experts in different countries, and obtaining the latest news on the nanotechnology profession. The INBN has participated in international fairs in Russia, Japan, Switzerland, and South Korea.

Table 6.2 Technology-Market Services Corridor (TMSC) Services

Technology-Market Services Corridor
Nanoscale Certification
Technology Readiness level
Documentation
Patenting
Technology Monitoring
Technology Transfer
Technology Guarantee Fund
Market Requirement and Monitoring
Legal Advisory
Business Planning
Production Advisory
Venture Capital
Standards and Certificates
Financing
Local Marketing
Global Marketing

In early 2010, the INBN was restructured and became the TMSC, based on the realization that the process of transforming ideas into products requires a wide range of information as well as managerial, legal, and financial infrastructures. The TMSC consists of service-providing firms and institutes active in the field of providing technology support, thus connecting the necessary links in the cycle of technology development. Through the TMSC, service-provider entities have been gathered into one centralized complex. Table 6.2 displays TMSC services.

When a nanotech firm wants to use TMSC services, first its technology is examined to make sure that it is in the nanoscale. Then, based on its needs, some of the services of TMSC are suggested to the firm. Costs of the services are subsidized by the INIC. The TMSC thus empowers companies to commercialize their products or technical knowledge. TMSC services include venture capital services such as sharing the risk with the investors, introducing investment opportunities, and supporting feasibility studies for investment projects.

It is necessary to mention that the INIC offers incentives for nanotech companies to locate in incubators and technology parks. Over the past decade, approximately 100 incubators and 20 technology parks in different parts of the country were established. The INIC preferred to encourage existing incubators to host and support nanotechnology businesses rather than establish new ones. Under a plan that was initiated in 2008, the Council evaluates and supports incubators that currently have nanotechnology units. In addition, units that join incubators are supported by the INIC in the development of their products. To encourage the use of nanotechnology within existing industries, the INIC has taken a range of measures collectively named the Penetration of Nanotechnology in Existing Industries Program. This program's actions include identifying industry needs, introducing potential applications of nanotechnology to various industries, supporting industrial managers in making visits to exhibitions, introducing recent nanotechnology achievements of Iranian companies to industry, and supporting feasibility study plans for the application of nanotechnology. The program began in 2008 by identifying various firms active in different industries and evaluating their approach toward the adoption of nanotechnology. To this end, relevant nanotechnology information is supplied to the companies, and if a company decides to use nanotechnology, the INIC will support the applicant company in the various stages of its implementation. As a result, currently around 40 companies from different industries are working on employing nanotechnology in their productive activities (INIC Annual Report 2008, 2009).

Iran Nanotechnology Laboratory Network (INLN)
Science and technology infrastructures are among the most important means for developing and promoting nanotechnology. The INIC tries to promote nanotechnology by providing research and industrial centers with the required financial support and facilities. In order to facilitate the provision of laboratory services for academic and industrial researchers, and to make better use of the country's laboratory potential, the INLN was founded by the INIC in early 2004.

The Network comprises a number of laboratories that possess the necessary equipment for analysis and characterization at nanoscale. These laboratories can provide nanotechnology researchers with services based on the network's directives. Currently, 44 laboratories from 12 provinces in the country are members of the Network, and two more laboratories are tentative members (INIC Annual Report 2010). Members of the network are evaluated on a yearly basis to identify the quality and quantity of the services they have provided to equipment users. According to the results of this evaluation, the INLN determines the amount of financial support for each member, and the memberships of those laboratories that do not meet minimum standards are revoked. There has been a significant growth (30 times increase) in the number of INLN users since its establishment.

In addition to providing financial support to member laboratories for purchasing and maintaining equipment and facilities, the INLN holds training workshops and courses for operators of laboratory equipment and enforces the international standard ISO/IEC17025 at the member centers (INIC Annual Report 2010).

To promote the production of instruments and equipment necessary for nanotechnology research and development, the INLN also runs a prepurchase program with equipment producer start-ups. The INLN finances laboratories to purchase domestically produced instruments, and in this way creates a market for the start-ups. Currently, there are 21 equipment and instrument producing companies that are capable of producing different nanotechnology equipment such as scanning tunneling microscopes (STMs).

The INLN's activities from 2004 to the present have shown that its member laboratories need a reference center fully equipped with standardized facilities, so that they can check their results through interlaboratory comparison and ascertain their accuracy. These activities have further shown the need for a center that could produce standard reference materials (SRMs), create a collection of such materials, and provide services to the members of the Network.

In addition, to monitor the nanotechnology product market so as to assure people of the quality and safety of such products, the country needed a specialized laboratory reference center that would analyze, approve, or reject nanoproducts. Therefore, the plan to establish the Center for Nanometrology and Development of Nanotechnology Laboratory Equipment was undertaken by the INIC. As this Center has a governing and supervising role, it is financed and managed by the INIC as the national entity in charge of this technology. The INIC is authorized to give NanoMark approval to the products that observe the standards set by the center. NanoMark on a product certifies that

- the scale of nanoparticles incorporated in the product is under 100 nanometers;
- the product has specific properties due to its nanoscale;
- the production process is repeatable.

Major Challenges of Nanotechnology Development

Iran's innovation process, like innovative activities in other resource-based economies, has different problems than those faced by developed counties, where innovations in most cases are market driven. Investment outlays in innovation projects often, but not always, result in profitability.[4] Of course, governments in these countries play a supporting role in the national system of innovation. For instance, there are many sector-specific or application-specific grant programs to promote business growth and collaboration with academic institutions and governmental labs in Organization for Economic Cooperation and Development (OECD) countries (Bahreini et al. 2010b).

In the Iranian economy, industry's demands for the production of better and newer products are often met by importing machinery and technology from abroad. Government investments intended to counter this trend by supporting research and development within Iran often do not result in improved or newer products. However, Iran has tried to change this paradigm recently, especially in the fields of emerging technologies, and has adopted a more market-driven approach. Government investments in such technologies are based on economic calculations and are designed to meet the needs of the industries.

Therefore, the main issue of Iran's innovation process is how research and development activities can best be integrated with business needs, and also how research results can be marketable so that a complete chain from science to wealth is created.

Investigation of nanotechnology companies' status in the value chain shows that they are faced with some challenges in how to commercialize their research outcomes: challenges that threaten the very survival of these firms. Under such conditions, focusing on the final parts of the innovation process, that is, commercialization and marketing, seems to be vitally important for the successful implementation of innovation policies in the country. In what follows, we will discuss some internal and external challenges faced by nanotech companies.

Nanotech Companies' Internal Challenges

The internal challenges faced by Iranian nanotech companies include inadequate technology, lack of business expertise, lack of focus, and lack of cooperative relationships with big industrial groups.

A significant challenge for nanotech companies is that the new technology embodied in the product or its manufacturing process can prove inadequate in some way. This can happen for a number of reasons:

- A product or process development has been technology driven rather than driven by customer needs.
- The R&D departments often make excessive claims about the product's potential value within the budget and time scales required by the company.

- Poor quality control has resulted from unfamiliarity with the process technology or a need to get to the market first (Meldrum and Millman 1991).

In one nanotech company, for example, extensive resources were devoted to the production of a new antibacterial and antiodor refrigerator filter. The basis for this outlay was the ability of the R&D team to develop the necessary technology for the product. However, the market needs were not fully considered in the development process. The result was that the company had to abandon the product, and thus wasted scarce resources.

In other cases, lack of quality control had resulted in claims by several companies that their products were substitutes for existing technologies, but the claims proved to be wrong in practice. For example, this situation occurred in the applications of nanosilver. Some companies claimed that nanosilver could be used as a substitute for some poisons or drugs, but results after applying the product did not confirm these claims. It was evident that the companies did not fully consider all aspects of the product nor complete necessary tests before making the claim. Eventually, this behavior caused a loss in customers' and stakeholders' trust in their products and harmed the commercialization process.

In spite of the fact that Iran now has many technical specialists in nanotechnology, there remains a lack of business expertise required for nanotech business development. Nanotech firms face several problems in planning their business model and commercialization activities, and often end up continuing on a trial-and-error basis.

Another challenge of Iranian nanotech companies is the temptation to work in diverse fields without enough focus and concentration on each field. For example, some of the nanomaterial producers tried to produce nanoenabled products themselves. Since new technologies bring new opportunities and fields of activities for the firms, some companies have tried to achieve economies of scope by becoming active in all those fields in order to achieve more benefits; however, because such technologies are extremely complex, they should give greater importance to specialization and proficiency. Focusing on one field and achieving profound competence in different aspects of the technology of the field could create competitive advantage for a firm.

The lack of interaction with established industries is another major obstacle facing Iranian nanotech companies. This lack of cooperation has caused firms to become separated from the real environment of industry and business. Such isolation decreases their power to identify and analyze growth and development opportunities. Research in this area has shown that successful high-tech businesses have formed close ties with industrial groups. These challenges might reduce the motivation of entrepreneurs and investors to enter the nanotech fields (Bahreini and Salehi 2010a).

Nanotech Companies' External Challenges

The major external challenges facing Iranian nanotechnology companies include a business environment that does not support the commercialization of technology and an institutional framework that does not contribute to the development of nanotech firms.

Necessary institutional infrastructure for the commercialization of nanotechnology in the country is still lacking. Mechanisms for obtaining licenses and required standards for products, securing financing, and protecting intellectual property rights are not formed yet. The commercialization path of a product faces the bottleneck of obtaining licenses and standards from Iranian regulatory agencies. These agencies are governmental organizations that usually lack information and the required expertise to meet the needs of applicants. Most of these agencies are incapable of serving nanotech firms, and the process is too time consuming for those who try to obtain required certificates. Therefore, firms cannot get the necessary permits to launch their products in the market at the right time (Bahreini and Salehi 2010b).

The activity of nanotech startups in a competitive business environment very much depends on supporting laws and policies that guide the development of these firms. Policy instability has caused managerial failures because of firms' long-term reliance on government backing. From the viewpoint of some successful Iranian nanotech companies, the sustainability of the government's policies is the most important category of governmental assistance programs. These companies consider the sustainability issue more important than the size of financial assistance (Saghafinia 2010).

Performance of the INIC

The performance of a technological development organization is measured by considering three indicators: human resources and knowledge creation, commercialization and finance, and outputs and markets (Enzing et al. 2007). These indicators for nanotechnology in Iran are discussed below.

Human Resources and Creation of New Knowledge

As mentioned above, two different support systems were used by the INIC to promote research and development in nanotechnology: horizontal and vertical support. Outcomes of these support policies are described below.

Outcomes of Horizontal Support Policies

Horizontal support policies that target researchers at universities and research centers have been put into practice since 2005. These supports included financial incentives for working on master's and PhD theses, scientific articles, books, and other publications. As a result, more than 60 Iranian universities and research institutes are currently involved in nanotechnology research. Statistics about nanotechnology scientific publications from February 2005 to December 2010 show that 1767 ISI articles, 5137 MSc theses, 1118 PhD dissertations, 803 international conference papers, and 79 books were published.

From a scientific point of view, Iran enjoys a high ranking in the field of nanotechnology among the countries of the world, and holds the premier position in the region. Taking the number of ISI publications as an indicator of science production, one can monitor Iran's improvement in this field. Since 2003, Iran has enjoyed the highest rate of growth in this ranking in the world.[5] The number of nano-related papers produced by Iranian authors and indexed by ISI in 2000 was only 8, and the number reached 1,845 by the end of October 2010. In the world ranking of ISI nano-related publications, Iran is among the top ten countries.

Among the Islamic Development Bank (IDB)-member countries that are pursuing scientific and engineering research related to nanotech, Iran accounted for approximately 29 percent of all the scientific nanotechnology-related output over the period between 2000 and 2008. Iran and Turkey had the majority of all publications, followed by Egypt and Malaysia (Islamic Development Bank Report 2009).

Outcomes of Vertical Support Policies

Vertical support policies aim at specific projects that concern the national priorities of the country. Table 6.3 shows the national priority of different nanotechnology fields, gives a brief description of each project, and lists its start date and current status. The number of projects that were supported through vertical policies of the INIC is very small compared to those supported through horizontal policies.

Table 6.3 Pilot Projects Run by the INIC and Their Current Status

No.	National Priority Field	Project	Start Date	Current Status
1	Energy	Producing insulator color for building	2008	Obtaining certificate and entering the market
2	Health	Early diagnosis of cancer	2005	Pilot stage/ obtaining certificates
3		Water purification at Karoon River using nanofiltration	2006	Pilot stage
4	Environment and Water	Producing membranes for nanofiltration	2008	Test stage
5		Desert stabilization using a nanotechnology-based solution	2009	Test stage
6		Mineral processing	2009	Research stage

Business Atmosphere

The INIC has tried to improve the business atmosphere for nanotech companies in various ways. Supporting incubators and establishing the TMSC and the INLN are examples of such activities that have already described here. The performance of the INIC on some other important issues like patents, employment, and venture capital investment is discussed in the following sections.

Patents

Since 2007, the INIC has supported patenting by financing up to 80 percent of the expenses for the registration of inventions with the most respected international bodies, such as the United States Patent and Trademark Office (USPTO) and the European Patent Office (EPO). Up to now, around 50 patents have been registered in internationally recognized patent registration offices abroad. For economy of space, we cite only a few examples of published patents below:

- A method of forming a carbon nanotube emitter with applications in nano-printing and use thereof;
- A process for hydroconversion of heavy hydrocarbonaceous feedstock;
- A catalyst based on MgO or alumina nanoporous support and preparation of the same;
- A carbon nanotube-supported cobalt catalyst for converting synthesis gas into hydrocarbons.

Employment

The INIC is very interested in increasing employment related to nanotechnology. Since 2009, it has initiated activities to help nanotechnology graduates find employment in nanotechnology fields and to encourage private companies and institutions to employ nanotech experts by funding approximately 50 percent of their salaries and benefits. The INIC also established a website to provide information about nanotechnology job opportunities (http://talent.nano.ir). Since 2009, there have been 94 requests to the INIC for admittance to the Employment Support Program, and as of January 2011, 72 of these applications had been examined and accepted. The program has supported 44 organizations that have employed MSc or PhD graduates of nanotechnology fields or the graduates of other fields that used nanotechnology concepts in their theses. The INIC envisages a proper growth in job opportunities and predicts an increasing number of active experts in nanotechnology.

Venture Capital Investment

The INIC has supported the establishment of some companies through venture capital firms as well. A total of 14 high-potential nanotech startup companies have been established and financed through venture capital investments. The total financing amounts to more than $4.5 million, of which approximately $3.7 million was invested (INIC Annual Report 2010).

Public Awareness

The promotion of nanotechnology and educating the public about this field are among the aims of a nanotechnology Future Strategy document. The efforts of the INIC in this area are briefly described below.

To raise public awareness about nanotechnology, the INIC publishes the *Nanotechnology Monthly Journal* with a circulation of nine thousand copies, and also sends the electronic version to 21,000 subscribers. The INIC website (www.nano.ir) provides information in four languages and has been very successful in attracting visitors and disseminating information.

With the purpose of promoting nanotechnology and supporting its commercialization, the INIC holds a yearly Nanotechnology Exhibition. Companies active in the field of nanotechnology, and universities and research centers present their latest products and achievements in their exhibits. The exhibition also has a "public show" section in which applications of nanotechnology in various fields are introduced to visitors in both theoretical and practical ways. The numbers of booths and visitors to the "Iran Nano" exhibition doubled in 2011 compared to the first exhibition in 2008.

Finally, the INIC established the Nano Club in order to promote students' knowledge of nanotechnology and facilitate students' research in this field. The Nano Club tries to direct students in an educational process that extends from learning the basic concepts of nanotechnology to creating knowledge-based ideas about the subject. The inclusion of nanotechnology basics in the high school curriculum has been one of the main activities of the Nano Club. In addition, compilations of scientific and educational books and CDs for students, the designing of games that help students understand nanostructures, and the organizing of student seminars, workshops, and exhibitions throughout the country are among the activities of the club.

International Cooperation

Success in the development of science and technology requires constructive interactions with domestic and international researchers on these subjects. For the IR (Islamic Republic) of Iran, which intends to become a key player in nanotechnology and to obtain at least 1 percent share of the global market for nanotechnology-related products, it is mandatory to have a constructive relationship with the entities that produce the remaining 99 percent of the market share. This objective is pursued through the formation of nanotechnology regional and international networks and cooperation with international organizations and other countries via joint projects, conferences, and training workshops. These international cooperative arrangements are discussed below.

ECO-NANO is a network comprising the Economic Cooperation Organization (ECO) member states, and it was formed in May 2009 through an initiative of the INIC. An intergovernmental organization involving seven Asian and three Eurasian nations,[6] ECO provides a platform to discuss methods to improve economic development, promote trade, and increase investment opportunities. The main objectives of the ECO-NANO network are to promote nanotechnology among the member countries, exchange experiences gained and knowledge acquired through R&D on nanotechnology by the member states, expand the economic share of ECO in knowledge-based commercial transactions, create regional and international nanotechnology markets, and finally, improve the standard of living in the member countries. The network was officially inaugurated on May 26, 2009. A two-year program was drawn up for its future activities, and a project on water purification using nanotechnology was included in the program.

In 2008, Iran proposed the formation of the International Center on Nanotechnology as a part of the United Nations Industrial Development Organization (UNIDO), with a focus on water and wastewater. In October 2009, the IR of Iran signed a document to inaugurate the center, which was placed under the supervision of UNIDO. This is UNIDO's first International Center on Nanotechnology, and its office is located in Tehran. It has a budget of $3 million over a period of four years, paid by the Iranian government (INIC Annual Report 2009).

In 2008, the INIC became a member of the Asia Nano Forum (ANF), and presented Iran's nanotechnology capabilities at the annual meeting of the Forum in Taiwan that October. The ANF, a network of 15 Asian and Oceanic countries, was founded in 2004 with the mission of promoting the research, development, and industrialization of nanotechnology among its member countries.

Other international affairs of the INIC include organizing a nanotechnology workshop composed of the science and technology centers of nonaligned movement (NAM) countries, organizing the second Iran-India joint conference on nanotechnology, and signing a number of memoranda of understanding with various countries such as China, the Ukraine, and Brazil (INIC Annual Report 2008 and 2009).

Summary and Conclusions

What policy lessons can we draw from a critical examination of nanotechnology development processes in Iran over the last decade? In this section, we synthesize these findings and present some recommendations for further improvement of technology policy decisions.

Unlike the country's previous experiences with technology development, Iran decided to pursue nanotechnology at a time when its development was a matter of choice and not a necessity. The country chose to develop the technology selectively as a vehicle for future innovation and technological change. Another significant feature of the nanotechnology development plan was that it was considered as a benchmark for the development of other technologies. A study of Iran's nanotechnology program that identifies its successful experiences and its mistakes and policy failures holds valuable lessons for policymakers and policy researchers.

Among the interesting lessons that can be learned from Iran's nanotechnology program is the effective use of promotional tools, and the creation of public and expert awareness about the subject. Advertisements about the importance of nanotechnology motivated many people, companies, and governmental organizations to invest in this field. University administrators granted special privileges for the academic programs dealing with the technology; different ministers put nanotechnology on their list of priorities; even some Provincial Governors wanted to invest in the nanotechnology area using government funds. This wave of interest, along with the support of public media, mobilized human and financial resources toward nanotechnology.

However, the INIC did not have attractive investment plans to offer to managers and companies, because investable opportunities in nanotechnology products were not developed at the time the technology was being promoted. This experience shows the effectiveness of the media campaign; however, it also indicates that timing was also a crucial factor for success.

An important aspect of Iran's nanotechnology program was that its designers were familiar with principles of technology policy and innovation systems, and they based the program on these theories, unlike previous technology development programs that did not have similar guideposts. They have also published their planning experiences in this field in books and articles.

Perhaps the most important element in technology policy planning is determining a proper target. From the beginning, the INIC set the target of placing Iran among the 15 countries most advanced in nanotechnology. The goal itself was appropriate, but because there was no clear indicator that could enable the policy makers to know whether the goal had been achieved, in practice the simplest indicator, namely the number of scientific articles, was chosen. According to this simple indicator, the target was met by the middle of the period that was established for the achievement of the goal. Nonetheless, more qualitative indicators like the level of technology and its impacts have not been used for evaluation yet. At this time, no specific nanotechnology and innovation indicators exist in the world.

Another important issue related to Iran's nanotechnology program is that since its inception, the program has faced unfavorable international conditions and sanctions against Iran. In confronting these sanctions, policy makers and technology experts alike have turned challenges into opportunities by developing indigenous technological capabilities that have increased self-sufficiency in many scientific fields, including nanotechnology.

The program has been in progress for five years and there are more expectations for it regarding revenue generation during the next five years. One might expect to find the measurement of total investment in the technology and the total value-added benefit this has provided to firms involved in nanotech activities to be one of the responsibilities of the INIC. Unfortunately, the INIC has not seriously considered the implementation of this important task. Calculation of this rate of return on the nation's investment should be the top priority of the INIC.

Although Iran's nanotechnology development program stresses the completion of all parts of the value chain, from science development to creating wealth, more attention has been paid to the first, less challenging part, which is development of a knowledge base of nanotechnology through academic research. As a result, the country has enjoyed great successes in training its scientific cadre, developing a knowledge-base on the subject, ensuring that the cadre publishes in reputable international scientific journals, and gaining valuable experiences in technology development. However, the level of achievement in commercializing the R&D results does not match that of scientific development. As a result, it seems that certain goals of the program were not met.

Perhaps the commercialization of R&D results requires maturity on the part of an industry, and industrial maturity is not a short-term phenomenon. Industrial development is a long-term process that involves not only one industry or one technology, but the synergistic interaction of the industrialization of many sectors of the economy, as well as the development of all institutions of the society. On the other hand, perhaps the technology planners did not properly implement thorough programs. It is well known, however, that as a result of the support by the INIC, several nanotech companies produced nano-enabled products. Nevertheless, the INIC should provide a suitable infrastructure for the development of other parts of the value chain, especially nanointermediate companies.

Notes

1. The views expressed here are those of the authors and do not necessarily reflect the views of the INIC.
2. Recently, TCO has been renamed the Center for Innovation and Technology Cooperation (CITC).
3. On entrepreneurship in Iran, see chapter 4 on the national innovation system in this book
4. As an example of massive R&D investment outlays by a large private enterprise that resulted in equally massive losses, we cite General Motors Corporation's $67.2 billion expenditures on R&D and net investment during 1980–1990, which resulted in GM's equity value of only $26.2 billion at the end of 1990. For details, see Jensen 1993.

5. See Maghrebi et al. (2011) for more information about the method of ranking countries.
6. ECO member countries are Afghanistan, Azerbaijan, Iran, Kazakhstan, Kyrgyzstan, Pakistan, Tajikistan, Turkey, Turkmenistan, and Uzbekistan.

References

Bahreini, M. A., and F. Salehi-Yazdi. 2010a. "Four Major Challenges of Technology Commercialization in Iranian Nanotech Companies." Paper presented at the 3rd International Congress on Nanoscience and Nanotechnology, Shiraz, Iran, November 9–11, 2010.
Bahreini, M. A., and F. Salehi-Yazdi. 2010b. "High-Tech Commercialization Challenges from a National Viewpoint: Case of Nanotechnology." Paper presented at the 4th National Conference on Management of Technology, Tehran, Iran. November 9–11, 2010.
Bahreini, M. A., F. Salehi-Yazdi, and Z. Abolhasani. 2012. "Descriptive Investigation of Nanotechnology Value Chain in Iran" (in Persian). *Journal of Science and Technology Policy* 3: 71–86.
Bahreini, M. A., F. Salehi-Yazdi, and M. Shadnam. 2010. "Comparative Study of Government Policies in Supporting Nano-tech Companies in Iran versus OECD Countries." Paper presented at the Conference on Commercialization of Micro and Nano Technologies, Santa Ana Pueblo, NM. August 29–September 1.
Enzing, C., A. van der Giessen, S. van der Molen, and G. Manicad. 2007. "BioPolis: Inventory and Analysis of National Public Policies That Stimulate Biotechnology Research, Its Exploitation and Commercialization by Industry in Europe in the period 2002–2005." BioPolis Final Report. Accessed August 25, 2012. http://ec.europa.eu/research/biosociety/pdf/biopolis-finalreport_en.pdf.
Fulekar, M. H. 2010. *Nanotechnology: Importance and Applications* New Delhi: I. K. International.
Ghazinoory, S., M. Abdi, and K. Bagheri. 2010. "Promoting Nanotechnology Patenting: A New Experience in National Innovation System of Iran." *Journal of Intellectual Property Rights* 15 (6): 464–473.
Ghazinoory, S., and M. Farazkish. 2010. "A Model of Technology Strategy Development for Iranian Nano-Composite Companies." *Technological and Economic Development of Economy* 16 (1): 25–42.
Ghazinoory, S., and R. Ghazinouri. 2009. Nanotechnology and Sociopolitical Modernity in Developing Countries: Case Study of Iran." *Technological and Economic Development of Economy* 15 (3): 395–417.
Iran Nanotechnology Initiative Council (INIC). 2005. The Future Strategy: The Islamic Republic of Iran's Ten-Year Nanotechnology Development Plan." Accessed April 9, 2011. http://en.nano.ir/files/futurestrategy_2006_2015.pdf.
———. 2008. *Annual Report.* Accessed August 25, 2012. http://www.nano-atu.ir/papers.php?s_ppr_id=114.
———. 2009. *Annual Report.* Accessed 25, 2012. http://www.nano-atu.ir/papers.php?s_ppr_id=123.

Iran Nanotechnology Initiative Council (INIC). (2010) *Annual Report*. Accessed August 25, 2012. http://www.nano.ir/paper.php?PaperCode=943.

Jensen, Michael. 1993. "The Modern Industrial Revolution, Exit, and the Failure of Internal Control Systems." *The Journal of Finance* 48 (3): 831–880.

Lux Research Inc. 2006. *The Nanotech Report: Investment Overview and Market Research for Nanotechnology*. 4th ed. vol. 1.New York: Lux Research Inc.

Maghrebi M., A. Abbasi, S. Amiri, R. Monsefi, and A. Harati. 2011. "A Collective and Abridged Lexical Query for Delineation of Nanotechnology Publications," *Scientometrics* 86(1): 15–25.

Roco, Mihail C., Chad A. Mirkin, and Mark C. Hersam. 2010. "Nanotechnology Research Directions for Societal Needs in 2020: Retrospective and Outlook." London: Springer.

Saghafinia, A. 2010. "Challenges and Opportunities Facing High-tech Companies: Case of ShafayeSari Co." Paper presented at the 1st International Conference on High-tech Export and Commercialization. Tehran, Iran. June 5, 2010.

United Nations, 2010 "Millennium Development Goals." UN.org. Accessed April 9, 2011. http://www.un.org/millenniumgoals.

US Environmental Protection Agency. 2007. "Nanotechnology White Paper." EPA.gov. Accessed August 25, 2012. http://www.epa.gov/osa/pdfs/nanotech/epa-nanotechnology-whitepaper-0207.pdf

Yahaya M., M. Salleh, I. Ho-Abdullah & Y. C. Chin 2009. "Roadmap for Achieving Excellence in Higher Education in Nanotechnology". Islamic Development Bank (IDB) Quick Win Project. Accessed August 25, 2012. http: //www.educationdev.net/ educationdev/Docs/Nano_Tech_Roadmap_Final_Draft.pdf.

CHAPTER 7

Biotechnology in Iran: A Study of the Structure and Functions of the Technology Innovation System

Tahereh Miremadi

Introduction

Many developing countries pursue a biotechnology based catch-up strategy in the hope of closing their technological gap with industrialized countries. Their strategies, however, are not similar, since biotechnology, which is a generic technology, encompasses several independent yet interrelated branches. Some countries have chosen to focus on just one branch, with the intention of becoming a specialized niche for the globalized market (on the case of stem cell research in Singapore, see Chaturvedi 2005); others have designed a comprehensive plan encompassing more or less all the branches. Iran as a member of the second group is a relatively vast country with a large variety of flora and fauna, biological reserves, and a population of roughly 76 million, which includes 3.8 million students in higher education.

Iran's development of biological sciences dates back to the 1920s when the Pasteur and Razi Institutes started to produce vaccines, Iran entered the modern biotechnology era in the mid-'80s with the establishment of the National Research Center for Genetic Engineering and Biotechnology (NRCGEB). Within a short time since then, the scientific community has built up significant capabilities in various related branches of biotechnology. The priorities set in the National Biotechnology policy document are biopharmaceutical, diagnosis, cell therapy, and regenerative medicine in medical biotechnology; biofertilizer, biopesticide, and molecular farming, in agricultural biotechnology; biosensors, biopolymers, and pharmacology, as well as industrial and mineral technology and waste management in the environmental sciences (High Council of Biotechnology 2005). From a statistical point of view, Iran ranks twenty-eighth in the world in terms of articles published, according to Scopus databank. Its rank among the Middle Eastern countries has been rising rapidly and, thanks to this accelerated growth rate, it finally surpassed Egypt and caught up with Turkey in 2011.

Currently, there are two thousand researchers and professors working in the research organization and universities. Around seventeen hundred students graduate from 50 universities and colleges annually in the country. The number of knowledge intensive start-up companies is 120 (Hajian 2011).

This initial progress attests to an underlying biotechnology sectoral innovation system in its formative phase, which supports the processes of biotechnology innovation and diffusion.

Based on the research literature (Mousavi et al. 2007; Ghareyazie 1999; Stone 2005; Moghaddam et al. 2011), we already know that the human capital is the most important strength of this system, and the public and the willingness of the government to support innovation in this field has been quite consistent. In terms of weaknesses, the system lacks sufficient up-to-date legal instruments such as biosafety laws and standard regulations. The system also lacks sufficient financing.

When the weaknesses and strengths are considered side by side, there seems to be a situation with some bottlenecks—in which consistent public and political support do not culminate in appropriate financial and legal environments.

This chapter aims to introduce these bottlenecks. However, it is very difficult to offer an in-depth analysis of the whole sector within the scope of one chapter. The complexity of the sector and the heterogeneity of the dynamics of the development of the branches of the sector in Iran (Baghai 2011) contribute to the difficulty of presenting the analysis in a single chapter. Therefore, we must focus on just one subsector: the field of stem cell research and technology, trying to address some missing function(s) in the biotechnology innovation system of Iran that causes obstruction and blockage in the system's functioning.

The Field of Stem Cell Research and Technology in Iran

When Iranian researchers registered a human stem cell (hSC) line at the International Society for Stem Cell Research (ISSCR) in 2003, it marked a milestone in the early phase of stem-cell research and development in Iran. Iranian scientists would follow this initial success with additional achievements in developing stem-cell research and technology in the fields of therapeutic cloning, cell therapy, and work with induced pluripotent (iPS) stem cells. The innovation system of stem cell research and technology (SCRTIS) is comprised of interconnecting actors, institutions, and networks that are undergoing a relatively outstanding formative phase. This chapter should shed light on various parts of this system, in part by answering these four basic questions:

1. What is the structural setting of stem cell research and technology in Iran?
2. What are the missing parts in the system, if there are any?
3. What prime movers drive emerging technological innovation system (TIS) systems in Iran?
4. What mechanisms facilitate and block this system?

This analysis uses a functional approach to TIS systems and observes that entrepreneurial activities were the driving force behind emerging technology system building in the formative phase. To give an empirical answer to the four questions above, this study has been divided into two different parts. The first provides a static profile of the TIS; the second undertakes a dynamic approach based on functional analysis and shows the cumulative causation between consecutive events leading to the accelerated process of system building in SCRT in Iran. The final part of this study integrates these findings to see what motor drives this technology and how some agents or mechanisms facilitate, block, or encumber the process of building the TIS system in Iran.

Theoretical Framework and Methodology for Collecting Data

The literature of innovation system identifies at least two basic approaches to identifying system failure.

One segment of the literature focuses on perceived weaknesses in the structural composition of a system. A central proposition in this approach is that the nature and composition of actors/markets, institutions, and networks may obstruct the formation of a TIS, and that eventually, such weaknesses in the system structure may lead to "system failure" (Woolthuisa et al. 2005; Lundavll et al. 2009).

Second, to offer a different perspective, some experts assert that in order to identify the central policy issues in a specific innovation system, one must supplement this structural review with a focus on the process itself (Bergek et al. 2008). They refer to this perspective as the "functional approach" and enumerate its components as follows.

The first function is the development and diffusion of knowledge (F1). The second is influencing the direction of a search, meaning the extent to which supply-side actors are induced to enter the TIS (F2). The third function is entrepreneurial experimentation, meaning, and knowledge development of a more tacit nature (F3). The fourth is market formation, that is, the articulation of demand in different markets, for example, niche market, mass market (F4). The fifth function is legitimation, which means the sociopolitical process of legitimacy formation through actions by various organizations and individuals (F5). The sixth function is resource mobilization, which indicates the extent to which the TIS is able to mobilize human capital, financial capital, and complementary assets from other sources (F6). Finally, the last function is the creation of positive externalities. It reflects the strength of the collective dimension of the innovation and diffusion process. It also indicates the dynamics of the system since externalities magnify the strength of the other functions.

This approach expects that actors, institutions, technologies, and networks will be arranged in a way that enables the system functions to work effectively, allowing a reasonable chance for the technology diffusion to increase. Therefore, since these components, in principle, should complement each other, they need to be put to work simultaneously to the greatest extent possible. The core of such analysis is to point out occurrences of positive feedback or cumulative causation in the process of system building. Recent TIS studies suggest that the direction of cumulative causation can be identified by pointing out interactions between system functions. Hekkert and Suurs (2007), in their analysis of biofuel, have shown a new method for operationalizing cumulative causation in innovation systems by starting with an event history analysis. This approach takes events themselves as the elementary units of analysis. This means that the unfolding of system functions over time is mapped in terms of events and sequences of events. Based on how these sequences map out, one can identify forms of cumulative causation and indicate how these influence the historical formation of TIS in the formative phase.

However, as Hekkert and Suurs suggest, there are different points of initiation for a TIS, which are determined by the driving forces behind it. The "motors of sustainable innovation," as the authors call them, encompass a science push motor, an entrepreneurial motor, a system-building motor, and market motors. Each one leads to a different path or itinerary to build a system (Hekkert and Suurs 2009).

To apply this theoretical framework to emerging stem cell research and TIS building in Iran, I needed access to a variety of data. The absence of a central data location in charge of collecting, classifying, and providing data in this field necessitated a variety of different strategies and multiple methods to increase the reliability and validity of the findings. First, I searched through media coverage of the subject at hand and then through internal archival documents of the Royan Institute, the major research institute active in this field in Iran. Afterwards, I did several interviews with the professionals associated with hSC technology in the country to confirm or refute my hypotheses. I also tracked publicly available information from the websites of the Iran medex and The Cell (the *Yakhteh* medical journal published by the Royan Institute) for scientific information; the *Howzeh* religious school; the website of prominent religious scholar Hojatollah Mobaleghi (the Director of the Islamic Science and Culture Academy [ISCA]), as well as two official websites of Iran's Supreme leader, Ayatollah Khamenei (the highest politicoreligious authority in the country)for official religious insights and discourse on the subject of the legitimacy of stem cell research.

PART I

Emergence of SCRT in Iran and Its Structural Map

The structure of the innovation system of SCRT in Iran stretches to encompass the actors (prominent research organizations, research hospitals, research universities, and private cell therapy clinics), networks (formal networks and strategic alliances in research and sharing data networks like the Iran Stem Cell Network, and the Iran Molecular Network) and regulatory, normative, and cognitive institutions, such as the *fatwa* issued by Ayatollah Khamenei, the Supreme Leader, in 2002 (Office of Leadership 2002). This *fatwa* gave researchers permission to use supernumerary human embryos in vitro (Gourabi 2011). In addition, other favorable bylaws as well as regulations on bioethics have been enacted in Iran during the last decade.

If we take a look at the structural map for stem cell research and development in the country, we observe a stem cell innovation system in its formative phase. There are public as well as private actors functioning as knowledge producers. The government-run bodies have set goals that are based on Iran's 20-Year-Vision Document, the highest policy document at the national level. The scientific networks coordinate the activities in the field of knowledge production and facilitate information sharing and scientific cooperation. Various clinics and hospitals perform different techniques of cell transplantation and therapy, while research institutions and universities produce knowledge and provide education.

Although in the real world, the interrelationship is not quite crystal clear, three interrelated and interconnected layers of governance, coordination, and performance for the production, diffusion, and application of technology can easily be distinguished in the stem cell research technology innovation system (SCRTIS) in Iran. At the level of governance, there are the Supreme Council of Cultural Revolution and the Supreme Council of Science, Research and Technology, which provide general guidelines for research policies. The Iranian Council of Stem Cell Research and technology (ICSCRT) under the auspices of the Vice-President for Science and Technology, established in 2009, is the governing body at the sector level. It is mandated to promote and finance research projects in universities and research organizations. The six special Committees of the Council deal with different issues, including intellectual property rights (IP), ethics, research, and so forth.

At the level of coordination, the scientific and knowledge networks are the major players in Iran. They function especially via the channels through which knowledge is diffused and shared. Stem cell technology in Iran has two formal knowledge networks: the Iranian Stem Cell Network and the Iranian Molecular Medicine Network, both of which were established by the Ministry of Health and Medical Education (MHME) to build a system of support for public-private partnership. The provision of financial support, capacity building, and the establishment of databanks are among their other announced missions. The establishment of new mouse embryonic stem cell (ESC) lines (2003–2004) is one of the accomplishments of these networks. However, the more effective and productive networks are informal ones formed on an ad-hoc basis among different universities and research and development (R&D) centers to perform joint programs, for example, the one established by the Royan Institute, the Small Business Development Centers (SBDC), and the Heart Research Center of Tehran University of Medical Sciencesto do autologous bone marrow stem cell transplantation into the infracted heart of a human.

At the level of the performance or the production and diffusion of technology, there are approximately 40 players in the country, both private and public, that can be divided into three categories: research universities, research organizations, and research hospitals. The big players here are incumbent organizations like the Royan Institute (an affiliate of the Academic Center for Education, Culture and Research [ACECR]), the Pasteur Institute (attached to the MHME), Shariati Hospital (run by Tehran University of Medical Sciences), and the universities in Tehran (University of Tehran and Tarbiat Modares University), and Shiraz University, Isfahan University, and Mashhad Universities in the provinces.

There are other players producing knowledge that are market-oriented newcomers. These new and energetic entities are start-up companies, which were established mostly by faculty members from the incumbent organizations. The most recognized ones are SABZ, established by a Tarbiat Modares University faculty member; the Cell Therapy Center of the Royan Institute; and the Stem Cell Research and Technology Company.

In viewing the role of institutions in system building, one must note that Iran does not currently have the necessary legal framework to deal with SCRT separately; therefore, the explicit legal instruments are mostly in the form of bylaws, which facilitates human embryonic stem-cell research by relying on the religious permissive *fatwa* (legal opinion of an Islamic jurist). In fact, the Iranian institutions have become favorable settings for research because of the religious authorities, who have legitimized SCRT and given their blessings to researchers who work on relevant projects (Office of the Leader 2005). Describing Islam as a science-friendly religion, the relevant literatures imply or maintain that the issuance of permissive *fatwas* stems from the view that some of the main sources of the Islamic tradition contain elements that support life science and technology innovations. On the other hand, the establishment of the Iranian Stem Cell Council by the Office of the Vice President in 2009 has paved the way for institutionalizing the facilitating mechanisms for technology innovation and diffusion. The interconnection among actors, networks, and institutions has led to accelerated growth in terms of scientific publications.

As discussed in the opening, Iranian researchers succeeded in registering a human embryonicstem-cell line in 2003. Building on this early success, they have had additional achievements in the development of SCRT, such as isolating ESC lines of six humans (Baharvand et al. 2004) and eight mice ESC lines (Baharvand and Matthaei 2004) and managing to convert these cells into functional pancreatic, heart, splenic, and liver cells. Other landmark achievements include coaxing human ESCs to become mature, creating insulin-producing cells in 2004 (Baharvand et al. 2006), cloning the country's first sheep in 2006 (Kazemi Ashtiani 2008), and conducting the world's first human ESC proteomics study (Baharvand et al. 2006). Moreover, two cell therapy clinics at the Royan Institute and the Hematology-Oncology and Stem Cell Transplantation Research Center are now active in curing such diseases as vitiligo and blood-related disorders, along with performing more than 35 clinical trials for nervous, gastrointestinal and hepatobiliary, cardiovascular, kidney and urinary tract, and musculoskeletal diseases.

In sum, Iran was the tenth country in the world to successfully generate human ESC lines in 2003 (Baharvand and Matthaei 2004; Baharvand et al. 2004), and the fifth to reprogram human skin cells to an embryonic-like state to create the so-called induced pluri-potent stem cells in 2009. In addition, the Iranian Hematology-Oncology and Stem Cell Transplantation Research Center ranks the third in the world when it comes to the number of total bone marrow transplantations performed in one organization (about four hundred in one year) (Ghavamzadeh et al. 2009).

PART II

Dynamic Analysis of SCRTIS

The findings of the first part of this chapter give a static profile of SCRTIS by which we can identify the structural components of Iran's innovation system. However, we still do not know the contribution of each component to the overall system goals or functions. This section will analyze the dynamics of the system to identify its functional pattern and locate the motor that drives the innovation system and moves it forward. That analysis will show that the Royan Institute activities entail far more than those usually associated with a research organization that produces and disseminates knowledge. After reviewing its contribution to SCRTIS building, one could easily recognize the Royan Institute as the entrepreneurial motor in this emerging TIS in Iran. This view is based on a careful consideration of the sequence of events and interpretation of the role played by the Royan Institute and other actors in the field. A historical review of TIS system building will show this more clearly.

Historical Review

In the short history of stem cell research and technology in Iran, the Royan Institute did not play the role of the pioneer. Instead, this credit goes to the Imam Khomeini Research Hospital, which, in a joint program with Shariati Hospital, began to perform bone marrow transplantations in 1982—a few years before the establishment of the Royan Institute. Moreover, Tarbiat Modares University (a university founded after the Revolution of 1979) was the first educational center that started to train and educate students in stem cell-related fields. However, it was the Royan Institute that introduced the application of stem cell research and technology in the framework of a new paradigm.

As a public, nongovernmental and nonprofit institute, the Royan Institute was established in 1991 in Tehran as a unit of the Iranian ACECR (*Jihad-e Daneshgahi* in Persian), along with the other revolutionary organizations under the auspices of the Supreme Council of Cultural Revolution (Miremadi 2010). It started its activity as a female fertility clinic. Progressively, it built its capacity through incremental innovation and interaction with universities, and obtained approval from the Ministry of Health as a Cell-Based Research Center in 1998. Its founder, the late Dr. Kazemi-Ashtyani (1954–2006), completed his studies at Tarbiat Modress University, and never attended any educational or training courses abroad. His colleagues early on were predominantly local graduates. However, Iranian alumni of biological and medical sciences from leading universities in Europe and the United States joined later.

When the Royan Institute entered the SCRT in 1998, it approached the matter in a whole new way. It galvanized and integrated the human capital of various knowledge-base interrelated disciplines, trying to close its technological gap with the other institutions, and eventually did overtake many research organizations in Iran and in the region.

The role of the Royan Institute as the entrepreneurial motor in the system could be briefly depicted as follows:

The Legitimizing Function (F5) and Social Entrepreneurship of the Royan Institute

If one applies the functional TIS model, one quickly sees that the Royan Institute was the major contributor to the Legitimizing Function (F5). The Royan Institute developed critical ties with the religious authorities, especially with the Office of the *Valy-e-faghih* (Juristconsult Guardian, or the Supreme Leader) and lobbied for the issuance of the permissive *fatwa*, which legitimized hSCRT use of supernumerary embryos. Additional actions taken by the Royan can be divided into three categories.

First, since innovations like stem cell research entail elements with implications that the key actors may not always fully comprehend, the crucial duty of the Royan Institute, as the "trusted" scientific authority, has been to help translate scientific jargon and assumptions into understandable material for religious authorities. Broadening comprehension of the scientific tenets and implications of stem cell research enables the religious authorities to focus on social demands for these innovations and the utility of novel technologies to meet them, while weighing at the same time the political value of the achievements as a source of pride for the country and the political establishment (Moballeghi 2007).

Second, the Royan Institute participates in the process of institution building by interacting with the Office of the *Valieh Faqih* and the *hozeh* religious seminary in the city of Qom. In so doing, the Institute benefits from the old network that its founder, a war veteran himself, built during the Iran-Iraq War with the cooperation of disabled soldiers (Mobaleghi 2007; Office of Leader 2008; ISNA 2006).

Third, the Royan steered the dialogue among different national stakeholders and helped construct a basis for national regulations governing stem cell research and its application (Mobaleghi 2007).

Such proactive involvement in setting the "rules of the game" requires social capital and suitable networks in which the entity is embedded, and a position that cannot be imitated or substituted by others. This is in accord with the argument byGulati (1999), who maintains that a firm's network relationship can create unique and non-substitutable value, and allow access to unique resources and inimitable capabilities of the other firms inside the network. This implies that networks bestow the firms with network resources, which are equivalent to the idea of "social capital" in an organizational setting. In the case at hand, the firm's stock of relational or social-capital network resources emanating from its prior relationship with the high-ranking political authority can be considered a very important resource. The key role played by the Royan Institute's networking is far more important than other institutional networking, because some nodes of the Royan's network consist of high-ranking political authorities. The Royan's engagement in institutional entrepreneurship built SCRTIS's legitimacy, which was crucial to the development of this nascent technology. The issuance of the aforementioned *fatwa* by the Supreme Leader paved the way for a wave of *fatwa*s issued by other religious authorities in the Qom *Hawzeh* seminary concerning issues related to animal cloning and transgenic animals, and so forth (Mohammadi 2008).

Apart from the religious aspect of SCRT, the legal and ethical aspects of this sector have also been an issue that the Royan has brought to the attention of its stakeholders. The Institute has strived in this regard by holding conferences and workshops in cooperation with the Law and Political faculties at the University of Tehran on such legal questions as prior informed consent and ethical questions relating to hSCRT. The interactions with the religious authority opened the door for raising public awareness of stem cell research, and the potential benefits it could bestow to the nation. Moreover, such research could meet political expectations of national pride and techno-nationalism.

Function of Knowledge Production (F2) and the Royan's Scientific Leadership

The issuance of the favorable *fatwas* helped to legitimize the research on hSCRT in Iran, and allowed the Royan Institute to conduct research projects in this field. Thisdevelopment led to a milestone achievement that was the registration of a stem cell line on behalf of the Royan in ISSCR in 2003. In 2004, the Royan Institute's Isfahan campus was founded and tasked with cloning and transgenic animal research (Mohammadi 2008). The outstanding scientific achievements include the establishment of the first organized Cord Blood Bank (2005); the first in-vitro maturation (IVM)-IVF sheep born in Iran (2006); the first cloned sheep (2006); the first report to the world of mesenchymal stem cells' new surface markers (2007); the first cloned goat (2009); the first cloned calves (2009); the establishment of the Royan Cell Therapy Center (2009), and the establishment of the Royan Stem Cell Bank (2011) (Jaenisch 2007; Morrison and Khademhosseini 2006; Pompe, Bader, and Tannert 2005; Royan Website, personal interviews).

Function of Resource Mobilization (F6) and the Royan's Financial Risk Taking

The social entrepreneurship of the Royan in the domain of institution building and knowledge producing led to high public expectations. At the beginning, some people seemed to hope that every disease could be treatable by stem cell transplantation and cell therapy. The Persian social media started thousands of websites and weblogs, praising scientific research in stem cells, in particular, and biotechnology in general. Astonishingly enough, many of them had a religious tone. In this regard, in an observational study that I did through the Internet in August 2011, I combined the word *Cellulhaye bonyadi*, for stem cell, with two different key words that have a politicoreligious overtone, namely, *shohadaye Jang*, for war martyrs, and *Defa'e Moghades*, for Sacred Defense (referring to the Iran-Iraq War), assuming that the contents of the scientific websites, which are normally neutral, do not contain these words. The search returned 296,000 hits, citing simultaneously "sacred defense" and "stem cell"; 13,300 hits citing "cloning" and "war martyrs"; and 928,000 hits containing "cloning" and "war martyrs." The other point of public interest in stem cell research is the publication of several books discussing human cloning through the Islamic prism (e.g., Mohammdi 2008) along with numerous conferences and seminars on the subject in Qom and Tehran.

The first research projects were financed on an ad-hoc basis by the Iranian Development and Renovation Organization (IDRO) within the framework of the SBDC. After several years of snowballing public awareness about the benefits of stem cell research, the annual national budget, in 2005, provided an independent line for the ACECR's program at the Center for Stem Cell Studies. This funding grew nearly ten times in the following year, and continued steadily to grow until 2008, when it fell to 16,000 million rials, equal to approximately $1.6 million (Government of IR of Iran 2010).

The third source of finance has come from the ICSCRT. Its annual financial support amounts $9 million.

Function of Market Formation (F4)

From the outset, the Royan Institute has been interested in initiating a plan for commercialization and financial mobilization. The aforementioned programs within the framework of the SBDC helped the Royan to traverse the gap between basic science and the more developed stage of clinical therapy, the commercialization of innovations, and the generation of income from the investment in stem cell R&D projects. It is expected that the number of spin-off companies will grow and function as a linkage between research and market. So far, one public cord blood cell bank and one bone marrow stem cell bank have been founded since 2005. The former has a network of 21 affiliates at the national levels. Moreover, the Royan Stem Cell Technology Company (Bon Yakhteh-e-Royan) was established to commercialize the lab regents like recombinant protein (basic-Fibroblast Growth factor) (bFGf). The commodity is designed and produced by the Royan research staff, taking advantage of a special credit line for governmental and private organizations offered by Leed Co., an affiliate of the IDRO, which is the big venture capital company that supports practical research projects with the aim of promoting science-based industries.

Table 7.1 Timetable of Major Events in the Course of SCRTIS Building

Year	Events	Role of the Royan
2002	Issuance of permissive *fatwa* on hSCRT	Institutional Entrepreneurship
2003	Registration of hSC line in ISSR	Technology Entrepreneurship
2005	Insertion of a budgetary line in the annual budget for SCRT program	Main beneficiary
2006	Tenfold increase in the GERD* for SCRT	Financial entrepreneurship: Main beneficiary
2007	Report of Mesenchymal stem cells' new surface markers to the world	Financial Entrepreneurship: Main beneficiary
2008	Birth of first cloned goat	Technology Entrepreneurship
2009	Establishment of Council for SCRT	First president of the Research Committee of the Council
2010	The emergence of multiple players	Provider of regents, cell lines, workshops, training

*Gross Expenditure on Research and Development.

Table 7.1 shows the chain of major events in the SCRTIS and the role of the Royan Institute in these events.

What Are the Facilitating and the Blocking Mechanisms?

The cumulative causation of the events shows there are two distinct phases in the system-building process of stem cell research and development in Iran. In the first phase, the Royan Institute plays the role of a single motor or prime mover for the system of innovation and the way it has been taking shape: It first paved the way for a scientific-religious dialogue between the religious and scientific communities that culminated in the issuing of the permissive *fatwa* for hSCRT in Iran through an intense networking with Qom religious authorities (F5). That led to a major technology development in the form of the production of the first Iranian hSC line that was registered in ISCCR [F2]. This event created, in turn, high public expectations for stem cell therapy to provide medical solutions for many diseases considered incurable. Rising expectations led to a wave of mobilization of resources [F6] and the entry of the Royan Institute into the phase of commercialization and helping start-up companies to produce more knowledge.

The second phase commences with the establishment of ICSCRT, which provides funds and incentives to the universities and research centers to diversify the knowledge base nationwide [F2].

However, there are already some symptoms that possibly indicate difficulties that could hinder the functioning of the system in the near future. To analyze these symptoms, one must return to the structural mapping and put the creation of three parts of the system, that is, performance, coordination, and governance, in a chronological order. The first part of the structural mapping that took shape was the performance layer in the 1990s, that is, the establishment of the Royan's Department of Stem Cell Research. The second part is the coordination that commenced almost in the period, 2000–2005, when the Stem Cell Network officially began its work. Finally, the part in charge of governance was the last part, which started in 2009 with the establishment of the Iranian Council of Stem Cell Research and Technology.

This sequence of events helps us better understand several points: first, it shows once again the fact that the single motor's entrepreneurial activities have contributed enormously to the formation of SCRTIS in Iran. This is because, apart from some nucleus in the Tarbiat Modares and Shariati Hospital, the Royan, as a research institute with a semi-nongovernmental organization (NGO) status, was the only organization active in the field of SCRT. Second, by the time other entities (universities and hospitals) started to work on SCRT, and in the absence of any coordinating and governing body, the Royan Institute's activities had outgrown the previous level of performance and extended earnestly to fill the organizational gap by financing the research and clinical trials of other organizations. In February 2012, out of about 35 registered clinical trials in the area of cell therapy, 30 were sponsored by the Royan Institute.

Although it is too early to evaluate the function of the ICSCRT as the governing body, a review of the scientific papers published in 2010–2011 indicates that the universities located in remote geographical locations from the capital, such as in Urumieh (Mohammadi et al. 2010), Kurdistan (Fathi et al.2011), Kerman (Seyed Jafari 2010), and Yazd (Khalili et al. 2011) are among the knowledge producers, attesting to the geographical diffusion of knowledge, to some extent [F3].

Many scholars suggest that the present system is an asymmetric system leaning toward performance and tending to be shaky and weak as far as governance is concerned. This problem hinders the system functions of the direction of research (F2), and might ultimately obstruct the process of the development of a technological innovation system.

The current situation does not favor a systematic approach and long-term planning for SCRT. As the policy document of SCRT has yet to be approved, there is no roadmap based on a consensus among the epistemic community and policy makers, since the decisions about research projects are made only by scientists, without contributions from strategic R&D managers and planners at the national level.

Conclusion

This study first structurally mapped the actors, networks, and institutions of the TIS of stem cell research in Iran and then evaluated the functional pattern of the system. It was submitted that below the surface of this static picture is a dual dynamism created by two major players, that is, the Royan Institute and the ICSCRT. The system-building process began with the Royan's contribution to the function of legitimization (F5). As a result, public expectations grew higher and higher after it made the scientific landmark event of registering the first Iranian ESC line and then continued on in a chain of successful scientific experiences (F1). The third major step was mobilization of resources (F6) and acquiring a governmental budget line. The supportive institutional environment and the Royan Institute's contributions and networking far beyond usual knowledge entrepreneurship are the facilitating mechanisms in the process of institution building and networking, and they reveal the Royan's role as the driving motor of system building. The Royan Institute has ended up creating significant scientific and intellectual capital in TIS, influencing the system's F1, F2, F5, and F6.

In the second phase, the momentum that the Royan's contributions have single-handedly created has been handed over to the ICSCRT. It has thus far encouraged some other crucial functions, including the entrepreneurial experimentation function (F3). Bearing in mind that the way to handle uncertainty is to ensure smooth entrepreneurial experimentation on a regular basis, we can understand the importance of this function, especially in the formative phase. These experimentations imply probing into new technologies and applications. The failure and success of these experimentations allow the unfolding of a social-learning process.

The process of entrepreneurial experimentations and the diversification of knowledge producers has created a knowledge-intensive niche market (F4) and fostered a dynamic user-producer relationship among the incumbent organizations that provide the cell lines, lab regents, and training to newcomers, which ultimately facilitates the availability of positive externalities (F7).

However, in order to maintain the current momentum in the future, the system needs to overhaul all its functions in the light of new global institutional and technological developments.

First, the existing regulations and bylaws do not cover all the current legal requirements of stem cell research according to international standards. While there are hundreds of patients with different medical conditions in the cell-therapy centers, the lawmakers are silent about how to license the cell-transplantation centers.

This fact shows that the function of F5 (legitimating) needs to go beyond the religious blessing. It is necessary to provide enacted laws and regulations explicitly covering the required technical conditions for the auditing and accreditation of the centers. Such legal instruments should be drafted in the process of cooperation among all stakeholders in tight-knit networks.

Obviously, in the absence of a comprehensive legal infrastructure, many opportunities for fundraising and resource mobilization (F6) are being squandered since excessive uncertainties about the legality of this type of research hinders the financial organizations' ability to support the research. For many scholars, the current level of budget is only 20 percent of what stem cell research in Iran really needs to maintain its momentum. The scanty budget makes the question of how to allocate this money to different options, a big dilemma for the steering committees of R&D centers since every choice made has huge opportunity costs.

By facilitating the functions of the resource mobilization and legitimating the cell transplantation (F5 & F6), the system is expected to have clearer views about the direction of research (F2). However, it must have been strengthened by different methods of technology forecasting and foresights with the collaboration of international organizations. Moreover, joining different international knowledge networks can help to align the domestic process of knowledge production with the global meta-trends at the international level.

The paper has identified a bottleneck in the biotechnology innovation system, which prevents public and political support from translating into the sufficient finance and appropriate legal instruments. Now, in conclusion, and after a systematic review of the dynamics of the field of stem cell research, it is clear that the system lacks a new level of (F2), that is, anticipatory and strategic governance as well as real-time monitoring by independent, impartial peer-review processes. This function does not belong to any specific player. In fact, there is no need to establish a new organization to remedy this situation; it is only a matter of commitment and cooperation among all the public-private players. For this function cannot be undertaken until the emergence of active participation by all the stakeholders, including the government, the public, religious institutions, legalists, planners, managers, and economists at the domestic and international levels.

References

Baghai, G. 2011. "National System of Innovation in Biotechnology in a Developing Country: A Gerschenkronian Approach to Biopharmaceuticals and Bioagriculture in Iran." PhD diss., University of Sussex.

Baharvand, H., S. K. Ashtiani, M. R. Valojerdi, A. Shahverdi, A. Taee, and D. Sabour. 2004. "Establishment and In-Vitro Differentiation of a New Embryonic Stem Cell Line from Human Blastocyst,"*Differentiation.*" 72 (5): 224–229.

Baharvand, H., H. Jafary, M. Massumi, and S. K. Ashtiani. 2006. "Development Generation of Insulin-Secreting Cells from Human Embryonic Stem Cells, *Development, Growth & Differentiation.*" 48 (5): 323–332.

Baharvand, H., and K. I. Matthaei. 2004. "Culture Condition Difference for Establishment of New Embryonic Stem Cell Lines from the C57BL/6 and BALB/c Mouse Strains."*In-Vitro Cellular & Developmental Biology* 40 (3): 76–81.

Barzegari A., M. A. Hejazi, N. Hosseinzadeh, S. Eslami, AghdamMehdizadeh, and M. S. Hejazi. 2010. "Dunaliella as an Attractive Candidate for Molecular Farming." *Molecular Biology Reports.* 37 (7): 3427–3430.

Bergek, A., S. Jacobsson, B. Carlsson, S. Lindmark, and A. Rickne. 2008. "Analyzing the Functional Dynamics of Technological Innovation Systems: A Scheme of Analysis." *Research Policy* 37: 407–429.

Chaturvedi, S. 2005. "Evolving a National System of Biotechnology Innovation: Some Evidence from Singapore." *Journal of Science, Technology and Society* 10 (1): 105–127.

Ghareyaize, B. 2000. "Iran: Hopes, Achievements, and Constraints in Agricultural Biotechnology." Proceedings of an International Conference, Washington DC, October 21–22, 1999. In *Agricultural Biotechnology and the Poor*, edited by G. A. Parsely and M. M. Lantin, 100–104. Washington DC: Consultative Group on International Agricultural Research.

Ghavamzadeh, A., K. Alimogaddam, M. Jahani, A. Mousavi, M. Iravani, B. Bahar, A. Khodabandeh, F. Khatami, F. Gaffari, and A. Jalali. 2009. "Stem Cell Transplantation: Iranian Experience." *Archive of Iranian Medicine* 12 (1): 69–72.

Gourabi, H. 2010. "How Did Royan Turn Out to be the "Royan?" (in Persian). Interview with Hamid Gourabi. The Center for Preserving and Publishing the Works of Grand Ayatollah Sayyid Ali Khamenei. September 7. Accessed August 19, 2011. http://farsi.khamenei.ir/audio-content?id=10065.

Government of IR of Iran. 2010. *The Budget Report* (in Farsi, *Ketab-e-Boodjeh*). Tehran: The Organization of National Management and Planning Publication.

Hajian, Zahra. 2012. "Requirements of Biotechnology Development in Iran." *Khorasan Newpaper*, no. 17794. 15. Khorasannews.com. February 12. http://www.khorasannews.com/News.aspx?id=635177&type=9&year=1389&month=12&day=22.

Hekkert, M. P., R. A. A. Suurs, S. O. Negro, S. Kuhlmann, R. E. H. M. Smits. 2007. "Functions of Innovation Systems: A New Approach for Analysing Technological Change." *Technological Forecasting & Social Change* 76: 1003–1020.

Iranian High Council of Biotechnology. 2007. "National Policy Document on Biotechnology Development: Designing the National Strategy for Biotechnology in Iran." Tehran: The Secretariat of Iranian High Council of Biotechnology.

Jaenisch, R. 2007. "An American Scientist in Tehran: Enthusiastic Stem-Cell Researchers in Iran Face Plentiful Funding but a Shortage of Equipment." Nature Reports Stem Cell. October 25. Nature Reports Stem Cell. Accessed August 19, 2011. http://www.nature.com/stemcells/2007/0710/071025/full/stemcells.2007.105.html.

Kazemi, Ashtiani S., M. H, Hasresfahani, S. M. Hosseini, F. Moulavi, M. Hajian, M. Forouzanfar, P. Abedi, M. Memar, M. Rezazadeh Valoujerdi, H. Gourabi, A. A. H. Shahverdi, H. Baharvand, A. Vosough Dizaj, H. Imani, P. Eftekhari Yazdi, M. Vojgani, M. Safahani, R. Radpour, I. Salahshouri. 2008. "Royana: Successful Experience in Cloning the Sheep." *Yakhteh Medical Journal* 10 (3): 193–200.

Lundvall, B.-A., K. J. Joseph, C. Chaminade, and J. Vang, J., eds. 2009. *Handbook of Innovation Systems and Developing Countries: Building Domestic Capabilities in a Global Setting.* Cheltenham, UK: Edward Elgar.

Miremadi, T. 2010. "Stem Cell Research and Technology in Iran-Window of Opportunity in the Midst of International Tension." *Review of Policy Research* 27 (6): 699–719.

Mobaleghi, A. 2009. "The Necessity of the Contribution of Jurisprudence in the Development of Technology Related to ART and Therapeutic Cloning" (in Persian). Qom: ISCA. July 24. Personal website of Ahmad Mobaleghi. Accessed August 19, 2011. http://www.moballeghi.ir/index.php/component/content/article /16 – /271.

Moghaddam A., and H. Hosseini. 2010. "Public Participation in Biosafety: What Should Be Done in Iran?" *Biotechnology Journal* 5 (3): 251–254.

Mohammdi, Ali, and E. Shabihsazi. 2008. (Human Cloning) Qom: Nashr-e- Moaraf.

Morrison, W. G., and A. Khademhosseini. 2006. "Stem Cell Science in Iran." Iranian Studies Group at MIT. October. Accessed August 19, 2011. http://isgmit.org/projects-storage/StemCell/stem_cell_iran.pdf.

Mousavi, Amir, Mohammed A. Malboobi, and Nasrin S. Esmailzadeh. 2007. "Development of Agricultural Biotechnology and Biosafety Regulations Used to Assess the Safety of Genetically Modified Crops in Iran." *Journal of AOAC International* 90 (5): 1513–1516.

Office of Leadership. 2002. "Text of *Fatwa* on Stem Cell Research." Tehran: Office of the Leader.

———. 2005. "Leader's Speech on His Visit to the Royan Institute." June 14. Center for Preserving and Publishing the Works of Grand Ayatollah Sayyid Ali Khamenei. Accessed August 19, 2011. http://www.leader.ir/langs/fa/?p=contentShow&id=3067.

Pompe, S., M. Bader, and C. Tannert. 2005. "Stem-Cell Research: The State of the Art." *European Molecular Biology Organization Reports* 6: 297–300.

Sanaei, M., and R. de Vries. 2008. "Embryonic Stem Cell Research in Iran: Status and Ethics." *Indian Journal of Medicine Ethics* 5 (4): 181–185.
Stone, R. 2005. "An Islamic Science Revolution?" *The Journal of Science* 309 (5742): 1802–1804.
Suurs, A. A. 2009. "Motors of Sustainable Innovation: Towards a Theory on the Dynamics of Technological Innovation Systems." PhD diss., Utrecht University. Accessed March 1, 2012. http://igitur-archive.library.uu.nl/dissertations/2009-0318-201903/suurs.pdf.
Suurs, A. A., O. Hekkert, and Roald Marko. 2009. "Cumulative Causation in the Formation of a Technological Innovation System: The Case of Biofuels in the Netherlands." *Technological Forecasting and Social Change* 76 (8): 1003–1020.
Woolthuisa, R. K., M. Lankhuizen, and V. Gilsing. 2005. "A System Failure Framework for Innovation Policy Design." *Technovation* 25: 609–619.

CHAPTER 8

Nuclear Technology: Progress in the Midst of Severe Sanctions

Behzad Soltani and Marzieh Shaverdi

Securing energy for the continuation of life is one of the most important problems of today's world. The declining sources of fossil-based energy, especially crude petroleum, have necessitated the production and use of alternative sources of energy. After the Second World War, the production of nuclear energy received the attention of many countries, so that by 2010, 441 nuclear power plants with 375-gigawatt electricity-producing capacity were active in 30 countries. Moreover, in 15 countries, 66 nuclear power plants with 63 gigawatts of capacity are under construction (IAEA 2011).

Today, it is considered a common practice to utilize nuclear reactors in securing energy, and many developed countries have included extensive use of nuclear power in their energy development plans. For example, the United States, with 104 reactors, has the most nuclear power plants. Other countries, such as France (58), Japan (54), Russia (32), South Korea (21), India (19), the United Kingdom (19), and Germany (17) also have nuclear power plants (IAEA 2011).

Based on such a widespread use of nuclear technology, it appears that acquiring the technology is a necessity, and nuclear technology provides valuable services to human societies. Iran, too, due to this fundamental need to secure energy, to follow common global practices, and to have political and economic independence, finds it necessary to construct and use nuclear power plants as well as apply nuclear technology for other peaceful purposes. Moreover, Iran's experiences have shown that if the preparatory work for securing fuel for its nuclear power plants from domestic sources does not take place, it may not be able to obtain it at all because global assurances for the provision of nuclear fuel by foreign producers, at least for Iran, could be unreliable (Secretariat of Science Cooperation Council of AEOI 2008).

The Islamic Republic (IR) of Iran, as a developing country, has found it imperative to develop nuclear power plants in order to achieve its strategic goals, including plans for the development of various sources of energy, energy security, acquisition of emerging technologies, as well as environmental, technical, and economic considerations, as part of its long-term plan for energy acquisition (Institute of Energy 2007).

In this chapter, we will review the history of nuclear technology development in Iran and the global standing of the country in this technology, analyze the national technology development documents, introduce the major actors in the field, and provide an overview of the future of the nuclear technology in the country.

The History of Nuclear Technology in Iran

The first attempt by Iran to acquire nuclear technology goes back to the 1950s. The steps the country took to secure nuclear energy were in collaboration with foreign countries, and most Iranian nuclear activities were in the framework of purchasing "turnkey" projects.

Ironically, the first country that encouraged Iran to acquire nuclear technology was the United States, which sold nuclear reactors to Iran for medical use and nuclear power generation.

A historical review of Iran's nuclear activities shows that the government of Iran signed a contract with AMF, an American company, for the construction of a research nuclear reactor at the University of Tehran (5 megawatts) in 1960. Accordingly, the first center for nuclear energy in Iran, the Tehran Nuclear Research Center, was established at the University of Tehran in the same year. At that time, Iran, a junior partner of the United States in the Central Treaty Organization (CENTO), had a close working relationship with this country, and all aspects of establishing the Center with respect to cooperation, selection of a foreign partner, and resolution of the technical problems were in accord with the United States' policies and interests (Documents & Diplomatic History Center 2007). Therefore, the Iranians had a marginal role to play in the construction and operations of the reactor. This was due to the American policy of not allowing the Iranian technical staff to be present in the area where the reactor was being installed. Of course, Iran did not receive any blueprints or technical information about the reactor from the Americans either. It should be noted that after installation of the reactor, the Iranian experts were fully engaged in the operations of the system, however.

The United States' decision to supply Iran with a research reactor for medical uses was based on President Dwight D. Eisenhower's "Atoms for Peace" program. In 1957, in the spirit of the program, which aimed to spread the peaceful use of nuclear technology in the United States and elsewhere in the world, the US government was instrumental in forming the International Atomic Energy Agency (IAEA) under the auspices of the United Nations. Iran signed the United Nations Nuclear Non-Proliferation Treaty (NPT) in 1968 (Alikhani 2006).

Based on the recommendation of the American government, the Atomic Energy Organization of Iran (AEOI) was established in 1972. Along with the establishment of this organization, the University of Tehran became active in teaching nuclear technology and began training students in this field. The University of Shiraz offered a degree program in nuclear engineering also. Moreover, Iranian students were sent abroad to study in the field of nuclear energy.

The year 1974 is considered a turning point in Iranian nuclear research. In this year, Iran concluded an agreement with Stanford University so that the Stanford Research Institute (SRI), one of the research arms of the university, could conduct a medium-term study of the prospects of socioeconomic and industrial development in Iran. The SRI produced a 20-volume report with the title "SRI Report," and conditioned economic and industrial development of the country on generating twenty thousand megawatts of electricity by 1995. The report concluded that the essential method of generating this quantity of electricity would be the construction of nuclear power plants in Iran (Alikhani 2006).

Based on the recommendation of the "SRI Report" for the generation of nuclear energy, on November 1974, Iran signed an agreement with Kraftwerk Union (KWU), a subsidiary of Siemens, to construct two light water reactors with the capacity of 1,300 megawatts. More than two thousand German and Iranian experts began working on this project, which at the time was one of the largest nuclear energy production plants in the world. The German company began work on the project in August 1975. This large project was supposed to be completed by 1980; however, it was left incomplete because of the Iranian Revolution and the Iraqi invasion of Iran in 1979.

During the 1970s, Iran concluded several more agreements with foreign companies to generate electricity through nuclear power plants. These agreements included a renewable contract with the United States to supply a 10-year nuclear fuel cycle to Iran (1974), with India in 1974, with West Germany in 1976, and with France in 1977 (Alikhani 2006). The European Gaseous Diffusion Uranium Enrichment Consortium (Eurodif) was a consortium formed by France, Belgium, Spain, Sweden, and Italy with the intention of constructing a power plant in the Tricastin Nuclear Power Center district in France. In 1975, Iran acquired a 10 percent interest in Eurodif after Sweden withdrew from this consortium and sold its interest to Iran. According to this agreement, Iran acquired the right to receive a certain quantity of enriched uranium from the factory for the production of radioisotope[1] medicine in the Iranian reactors (Documents & Diplomatic History Center 2007).

Overall, Iran invested $2 billion in Eurodif. One billion dollars was the share of Iran's direct investment in the project, and the second billion dollars was provided to the consortium as a loan.

Regarding another nuclear power plant venture, after much competition, the British and French reached an agreement so that these countries could start a joint study on establishing nuclear facilities to advance the progress of the nuclear fuel cycle in Isfahan in 1976.

France formally began its nuclear activities in Iran in 1977. Iran and Framatome, a French company, concluded an agreement in October 1977, for the construction of two nuclear power plants with a 900-megawatt capacity near Ahvaz, a city in the south of Iran. In December 1977, KWU also concluded a $4.8 billion contract to build 4 reactors in Iran (Documents & Diplomatic History Center 2007).

The shah's pursuit of nuclear technology created a great deal of anxiety among Israeli and Western leaders about Iran's potential to acquire access to this technology. The nervousness of the Israelis and the Western powers led them to exert pressure on the Western nuclear power companies to abandon their activities in Iran, with the result that there were considerable delays in the construction of nuclear power plants, which ultimately led to the cancellation of agreements after the triumph of the Revolution in 1979. Nevertheless, about 85 percent of the first reactor and 65 percent of the second reactor in Bushehr were completed just before the start of the uprising in February 1979 (Alikhani 2006).

On the other hand, realizing the Israeli lobby's role in encouraging the Western countries to nullify their nuclear power plants agreements with Iran, the shah turned to nonwestern countries and signed a secret agreement with South Africa in a sizeable deal to purchase nuclear equipment and raw materials.

During the revolution of 1978–1979, most German nuclear experts left Iran, and work on the project was halted. After the Revolution, the Iranian government cancelled the agreement to finish construction of the power plant. However, officials decided to resume building the Bushehr power plant, although by now, the Western countries' previous policy of technological cooperation with Iran had transformed into a policy of sanctions and a blockade of technology transfer. KWU was approached to resume the work under the terms of the agreement; however, it refused to do so, and offered that instead of nuclear fuel, the power plant should use natural gas turbines to generate electricity.

Iran refused the recommendation based on two economic calculations. First, Iran reasoned that using natural gas to generate electricity would have a substantial opportunity cost, because burning this very scarce resource would not permit it to be used in making petrochemicals, products that would contribute much value-added to the national output. Second, the country's known uranium ore reserve is estimated to generate a quantity of electricity that is equivalent to using 45 billion barrels of petroleum, which is roughly about 50 percent of Iran's known petroleum reserve of 96 billion barrels (Sahimi 2005).

Iran asked the German government to permit KWU to ship the reactor components and technical documentation under the International Chamber of Commerce (ICC) ruling of 1982. The ICC ruling had ordered the German company to deliver to Iran all materials and components for the nuclear power plant stored outside the country. The German government refused to do so, and Iran took the matter to the ICC in August 1996, and requested a compensation of $5.4 billion for the German company's failure to comply with the ruling (Koch and Wolf 1997). In spite of the Germans' receiving 85 percent of the cost of the project, failing to complete the work, and refusing to ship the main equipment and parts, the final ruling of the ICC was in favor of the German company (Alikhani 2006).

After the 1979 Revolution, with the unilateral termination of the contract by Eurodif and the refusal of the French government to refund Iranian investment funds, Iran sustained heavy losses. With the termination of the contract, Eurodif went to the arbitration court of the International Chamber of Commerce in Paris with a complaint against the AEOI. The court ordered all Iranian assets in Eurodif be frozen. After ten years of negotiations between Iran and France over the refusal of the French government to permit the company to comply with the terms of the agreement with Iran, 900 million French francs out of $2 billion of the Iranian investment was retained by the French company, and the balance of the invested fund was returned to Iran in the form of Iranian purchases of French products.

Following the Revolution, the most important partner of Iran in nuclear technology was the People's Republic of China. Iran entered into negotiations with China toward the middle of the 1980s, and reached important agreements with her for the exploration and extraction of uranium there. For example, in 1992, Iran came to a $110 million agreement with China for construction of a uranium conversion facility (UCF) site in Isfahan. In spite of this agreement and due to American pressure, the Chinese were unable to continue their cooperation with Iran (Mohammadi 2011).

The Iraqi military attacks against Iran, in addition to inflicting heavy damage on unfinished nuclear infrastructure, stopped nuclear activities in the country altogether during the eight-year war. During the war, while facing acute shortages of energy, Iran approached Japan and Spain about completing the nuclear power plant in Bushehr. However, due to American intervention, these countries refused to cooperate with Iran.

After the end of the Iran-Iraq war, the Iranian government developed various plans for the acquisition of nuclear technology and the production of nuclear energy. Among the country's nuclear achievements of the postwar years we can cite the completion of the Bushehr nuclear power station in cooperation with the Russian Federation, the finishing and expanding of nuclear energy facilities in Isfahan, and the establishing of uranium enrichment facilities in Natanz (Alikhani 2006).

During the administration of George W. Bush, the United States began an extensive effort to stop Iran's peaceful nuclear activities. Under pressure from the Israeli lobby in the United States and in Western European countries, the IAEA referred Iran's nuclear dossier to the United Nations Security Council for alleged violations of the terms of the NPT in September 2003. Toward the end of that year, the foreign ministers of France, Germany, and the United Kingdom visited Tehran to discuss their concerns about Iranian nuclear issues. During these negotiations, Iran agreed to suspend its nuclear activities, and voluntarily signed an Additional Non-Proliferation Protocol without being obligated under the terms of NPT to do so. With the signing of this agreement, Iran aimed to eliminate any misperceptions concerning its nuclear program. The European signatories of the "Tehran Accord," in turn, agreed to first explicitly recognize Iran's rights to conduct nuclear research, and second to discuss methods by which Iran could provide "satisfactory assurances" about the peaceful nature of its nuclear research programs. After Iran met these requirements, the European countries had to provide Iran with easy access to advanced nuclear technology.

However, as a result of the European powers' delay in providing Iran with access to nuclear technology, and their adoption of a policy of marking time to impede Iranian research and development (R&D) of nuclear technology, Iran in February 2005 requested that the European negotiators speed up the negotiation process, a request that was refused. Facing European intransigence, and shortly after the election of President Mahmud Ahmadinejad, Iran had no alternative but to resume its nuclear activities by breaking the IAEA's seals on the equipment in the UCF nuclear facilities in Isfahan, in August 2005. Thus, a new era of Iranian R&D in nuclear technology began.

In September 2005, the Board of Governors of IAEA, based on its 2003 report, declared that Iran had not complied with its safeguard agreement and voted to report Iran to the UN Security Council. With the sponsorship of the United Kingdom, France, and Germany, and the backing of the United States, the vote set the stage for a number of Security Council resolutions against Iran. In response, Iran suspended all voluntary cooperation beyond that which the country was obligated to observe with the IAEA, including implementation of the Additional Protocol (Alikhani 2006).

On April 8, 2006, President Ahmadinejad officially announced Iran's new technological capabilities in enriching uranium, as well as the establishment of a complete chain of uranium enrichment centrifuges in Natanz. This day, by decree of the Supreme Council of Cultural Revolution, was declared to be the official National Day of Nuclear Technology (Alikhani 2006).

Nuclear Energy Agreements and Contracts

A review of Iran's contracts and agreements with other countries with respect to the acquisition of nuclear energy prior to the revolution of 1979 and establishment of the Islamic Republic (IR) of Iran shows that these agreements were in the form of purchasing turnkey technology rather than the transfer of technology. Of course, one should not ignore the educational benefits to the employees who worked on these turnkey projects (Targhi 2001).

Iran had nuclear cooperation agreements with the following countries: the United States (1958), Canada (1972), France (1975, 1978), Germany (1974, 1976), India (1976), the Russian Federation (1993, 2006), and China (1993). The terms of these agreements regarding scientific and industrial cooperation consist of the following:

1. Exchange of information concerning the design, construction, and use of research reactors and their utilization in research and radiotherapy treatment (United States)

2. Use of radioactive isotopes in research in physics and biology, radiotherapy treatments, agriculture, and industry (Germany and United States)
3. Cooperation on health issues and the safety of operations at research reactors (United States)
4. Transfer of nuclear technology to Iran with emphasis on the right to transfer inventions, blueprints, and designs and to exchange scientists and engineers (Canada)
5. Extensive cooperation in science, technology, and industry, and the training and offering of scholarships to Iranian experts (France, India, Russia, Germany, and China)
6. Cooperation in the construction of the Uranium Conversion Facility in Isfahan (China)
7. The right to procure any nuclear-related equipment, instruments, and facilities (France).
8. Establishment of a center for nuclear research and development in Iran (France)
9. Implementation of the required steps to construct a nuclear power plant with a capacity to generate five thousand megawatts of electricity in Iran (France)
10. Cooperation in the desalination of sea water and the turnkey delivery of two nuclear power plants in Iran (France)
11. Exchange of information, scientists, researchers, technical personnel, and research results, and the provision of consulting services by Germany to Iran (Germany)
12. Design, construction, and management of nuclear power plants and other nuclear research facilities (Germany, France, and Russian Federation)
13. Provision of the safety and security of nuclear facilities and the prevention of the dissemination of the radioactive materials in the environment (Germany)
14. The use of nuclear energy for applications other than electric power generation (Germany)
15. Scientific and applied research concerning the security of the nuclear power plants (Russia)
16. Provision of consulting services in research and technology (Russia, China)
17. Formation of collaborative working groups for the implementation of specific projects dealing with development of scientific research and technology (Russia)

The Main Players in Nuclear Technology Development in Iran

The most important active institutional players in the Sectoral Innovation System (SIS) of nuclear technology in Iran are the following entities.

The Government

One of the main tasks of the government is to formulate general policies for the nuclear industry. The government also provides the financial resources for the development of the industry. Policies relating to the nuclear industry are formulated by the Supreme National Security Council, which consists of the AEOI, the Ministry of Foreign Affairs, and national security and related organizations.

Atomic Entergy Organization of Iran (AEOI)

The AEOI is responsible for the development of nuclear technology in the country. The law authorizing formation of the organization was passed four years before the Islamic Revolution, in 1975. Among its main responsibilities, are creating a scientific and technical infrastructure for the use of nuclear technology in economic development; conducting nuclear research and applying nuclear technology in the agricultural, industrial, and service sectors of the economy; generating electricity through the use of nuclear energy; cooperating with other countries on nuclear technology; representing Iran in the United Nations IAEA; and attracting foreign investment to the scientific, industrial, and mining sectors that are involved in nuclear activities. Furthermore, to achieve its goal, the AEOI is authorized by law to invest and open offices abroad, conclude cooperative agreements with foreign individuals and enterprises, or employ the needed foreign nuclear technology experts. Finally, the organization is authorized by law to purchase materials, instruments, and equipment from abroad in those cases where these items are not available in Iran, without the organization being required to secure any import permits from other Iranian government agencies (Targhi 2001).

The Universities

In the 2011–2012 academic year, 58 universities offered bachelor degrees, 25 universities master's degrees, and 10 universities doctoral degrees in nuclear physics and nuclear engineering in Iran. In the same academic year, 28 students enrolled in the bachelor program in nuclear medicine in the country. During this same academic year, 1,880 students enrolled in undergraduate programs, 240 students in master's programs, and 25 students in the doctoral programs in nuclear physics in Iran (National Education Assessment Organization 2011).

In 2011, 102 students graduated in nuclear engineering, and 277 students graduated in nuclear physics (The Office of Higher Education 2011).

The curriculum of nuclear engineering in the Iranian public universities consists of subjects such as nuclear medicine and reactor- and fuel cycle engineering. The enrollment in different nuclear engineering master's programs in the 2011–2012 academic year consisted of 53 students in radiation applications, 16 students in nuclear fusion, 26 students in nuclear medicine, 70 students in reactor design, and 29 students in fuel cycles engineering (National Education Assessment Organization 2011). Moreover, 24 students enrolled in the nuclear engineering doctoral programs at the public universities in the 2011–2012 academic year (National Education Assessment Organization 2011).

The Islamic Azad University of Iran, a private university, also accepts students in nuclear physics and nuclear engineering at the master's and doctoral levels. In 2010, this university accepted more than 700 students at the master's level in nuclear physics and nuclear engineering.

Given that the Iranian universities have the capacity to educate more engineers and scientists in the field of nuclear engineering than meet the country's manpower requirement in the nuclear technology fields, Iran has the capacity to supply scientists and engineers skilled in these areas to other countries.

It is important to note that Iranian nuclear experts are active in two professional organizations: the Nuclear Science Association and the Nuclear Medicine Association. The Nuclear Science and Technology Research Center (NSTR), in collaboration with the nuclear associations, publishes the *Journal of Nuclear Science and Technology*. This journal was established in 1985, and has published 56 issues thus far. Moreover, at least two national or international conferences on the subject of nuclear science and technology are organized in Iran annually. Finally, it should be noted that the Iranian universities play a relatively minor role in nuclear research, and most research activities in this field take place at the research institutes.

Iran has faced global constraints in training human resources in nuclear physics and nuclear engineering. However, these restrictions had little or no adverse effect on the growth of nuclear and other technologies in the country. In fact, it appears that these limitations have stimulated the domestic training of scientists and engineers. Iran has premium institutions of higher learning that are capable of training high-ranking scientists and engineers who can provide engineering services not only for Iran but also to other countries. Moreover, in terms of expertise, Iran enjoys a favorable global position, and some of the countries in the region have shown interest in receiving assistance on these subjects from the Iranian universities (Aghamiri 2010).

To counteract the potential retarding effect of the sanctions imposed on Iran, the number of skilled individuals in this and other technologies has increased. Faculty members and graduate students in nuclear physics and nuclear engineering have had major achievements, and these are manifested in their scientific publications, which reached three hundred articles published in reputable scientific journals in 2010 (Aghamiri 2010).

Nuclear Science and Technology Research Center

The Center is the research arm of the AEOI. Two main goals of NSTR consist of "development and expansion of research on nuclear science and technology" and "laying the groundwork for improvement of research in this field" (Secretariat of Scientific Cooperation Council of AEOI, 2008).

In 2011, with a three-hundred-member scientific staff and two thousand employees, this Center accepted 12 doctoral students in the nuclear science and engineering fields. The Center has eight research institutes: Physics of Plasma and Nuclear Fusion; Laser and Optics; Reactors and Accelerators; Agricultural, Medical, and Industrial Radiation Applications; Fuel Cycle; Nuclear Science; and Materials.

Currently NSTR owns and operates one gamma-ray irradiation system that is used in the sterilization of medical instruments. Most of the radiopharmaceuticals in the country are developed and produced in this Center.

Companies That Produce Parts and Components

More than three hundred private firms, both domestic and foreign,[2] are active in the production and acquisition of parts and components for the AEOI. Most of the high-tech components are manufactured inside Iran. In fact, the economic and technological sanctions by the Western countries have contributed to the rapid development of technological self-sufficiency in many areas of nuclear technology, such as fuel cycle, gamma irradiation, and accelerators, and the production of upward of 80 percent of the total parts in many instruments inside the country.

Analysis of the Nuclear Innovation System in Iran

The strength of the nuclear industry of Iran rests in the active participation of all the major players of the system in the innovative processes. The policy-making centers, research and development centers, investors, and production units promote and develop the industry in the country. Even though the government is the main user of the nuclear technology, particularly the Bushehr power plant and fuel cycle in the country at this time, the industry has private customers, especially the medical establishments that use nuclear medicine and nuclear technology in the sterilization of medical instruments. Moreover, through the expansion of irradiation systems for agricultural products, and the commercialization of nuclear byproducts such as depleted water for medical applications, and the development of industrial uses of zirconium, the private sector's use of the technology is on the rise (Soltani 2011).

Thus far, the goals of the nuclear industry are threefold: political, technological, and commercial. In the political domain, the country aims at safeguarding and promoting its national pride, developing technological independence, satisfying popular demand in Iran for the acquisition of nuclear technology, and above all, resisting the unreasonable dictates of foreign powers and manifesting a concrete display of national will and resolve to develop the technology irrespective of the present and potential costs. After over three decades of sanctions, the daily threat of military attacks against the country, and the creation of all sundry forms of impediments against the technological progress of Iran, the development of nuclear technology no longer has a one-dimensional goal of technological progress. Now, for most Iranians, technological progress, especially in nuclear technology, is a symbol of national resistance against foreign meddling in Iranian domestic affairs.

The export of nuclear products and services to Middle Eastern and Central Asian countries is another goal of the decision makers. A quarter century of Iranian experience in this field compared to that of neighboring countries opens up opportunities for countries that own advanced nuclear technology in cooperation with Iran to increase their presence in the markets for the peaceful use of nuclear technology in this region.

One of the weakest points of the nuclear innovation system in Iran is the inadequate coordination of activities among the main players in the industry. This weakness particularly manifests itself in nuclear R&D. It was possible to perceive the AEOI in the past as an agency in which there was no coordination in the R&D activities of its units. These commercial and research units, in pursuing the needs of their individual technological units, acted independently of each other. However, recent reforms in creating uniformity in the technological requirements of the constituent research units and companies of the AEOI, the research activities of the AEOI are organized according to the fields of nuclear technology. The present structure allows for more effective interactions among the units, and better understanding provides a suitable direction for the research activities and meets the industrial needs of the units.

Other weaknesses of the nuclear innovation system in Iran are difficulties in forming markets for nuclear technology, the need for increased innovation, and the need to enhance entrepreneurship in the industry. To overcome these shortcomings, it is necessary to develop plans for establishing a center for innovation in nuclear technology and assist in the formation of small and medium-size companies in the field so that the peripheral needs of the nuclear industry are satisfied.

National Plans and Documents for Development of the Technology

The passage of the Acquisition of Peaceful Nuclear Technology Law, the content of which was first adopted by the Council of Atomic Energy of Iran in 2003, and which was later passed by the Iranian Parliament in June 2006, forms the legal basis of the country's activities for the acquisition of technical capabilities in nuclear technology, as well as the production of nuclear energy for peaceful purposes. All of the subsequent planning documents in development of nuclear technology are based on this law (Institute of Energy 2007).

According to all technological and economic development plans of the country, i.e. Vision 1404, the Fourth (2007–2011) and Fifth (2011–2015) Development Plans, acquisition of electricity via nuclear power plants is considered one of the top goals by the planners (Institute of Energy 2007). All of these documents require that advanced knowledge and technology be attained in order to design, construct, and utilize nuclear power plants for the production of electricity. Moreover, international interactions and the mobilizing of resources to expand the country's scientific production in the nuclear field are set as other goals (Iran Expediency Council 2003).

The executive branch of the government is responsible for development of a general plan for the protection of nuclear power plants no later than the end of the first year of the Fifth Development Plan (2012) for approval by the Cabinet of the President of the Republic. Moreover, the executive branch is assigned the following tasks, which are to be achieved during the Fifth Socio-Economic Development Plan:

1. Development of a 20-year plan for generation of nuclear power.
2. Identification of new locations for the construction of nuclear power plants.
3. Attracting domestic and foreign capital for investment in the construction of nuclear power plants.
4. Planning for the procurement of fuel for nuclear power plants in Iran, and the construction of research reactors to produce radiopharmaceuticals and use in agricultural and industrial research.
5. Development of a general plan for the exploration of uranium ore and the acquisition of raw materials for nuclear fuel in Iran.

The Global Position of Iranian Nuclear Technology

The Iranian nuclear technology development program began in 1974, when the consumption of electricity in the country was approximately 3,500 megawatts, and the country had a population of thirty million. At that time, the needed electricity for a 20-year development plan was estimated to be 20,000 megawatts. This plan was supported by the United States in the 1970s. Now, with a 30,000-megawatt demand for electricity in the country, which has a population of 76 million, the planned generation of electricity using nuclear energy over the next 20 years is only 7,000 megawatts (Karkhanehei 2010). This implies that the energy needs of Iran are substantially higher and the current planned production of power is considerably lower than what were estimated in the 1970s. In short, the rising demand for energy makes a compelling case for the acquisition of nuclear technology.

Nuclear technology is among a limited number of domestically developed technologies in the country. According to IAEA's report, Iran is ranked tenth in the world in nuclear technology in spite of serious opposition by Western countries, in particular the United States, to Iran's acquisition of the technology (Karkhanehei 2010).

Technical Discussions of Nuclear Energy Production

Nuclear technology has been used in many applications, including the generation of electricity, the production of pharmaceuticals, nuclear medicine, the irradiation of food and health products, and in agriculture. However, the most notable and controversial application of nuclear technology is in electricity power generation. To gain a better understanding of the physics of nuclear power plants, we turn next to a discussion of the nuclear fission process.

An atom's nucleus can be split apart, and the word "fission" means to split apart. Nuclear fission releases a large amount of energy.

Nuclear power plants with thermal electric generating systems are distinguished by the methods of heat production. A nuclear power plant uses uranium as a "fuel," and uses the fission process to change the energy of atoms into heat energy. After mining, uranium is processed into tiny pellets that are loaded into rods and then put into the power plant's reactor. Inside the reactor, uranium atoms are split apart in a controlled chain reaction. This chain reaction releases heat energy that is used to boil water in the core of the reactor. This water is sent to another section of the power plant. In the heat exchanger, it heats another set of pipes filled with water to make steam. This steam turns a turbine to generate electricity.

Iran is involved in the R&D of peaceful applications of nuclear technology and has made major achievements in developing domestic technological capabilities in this field. We discuss these achievements briefly below.

Design of Medium-Size Power Plant

The design of the nuclear light water power plant IR360, in the city of Darkhovein (in southern Iran), with a 360-megawatt capacity, which fully utilized the engineering capacity of the country, began in the early months of 2005. The conceptual design of this project began in June 2005, and after the necessary analysis, it received approval in early 2006. At the time of the writing of this chapter in March 2012, the work on the complete, detailed design of the project with the aim of the final design and technical specification facilities, is underway. The design process is estimated to be completed in 2016.

Uranium Enrichment

In April 2005, Iranian experts were able to enrich uranium in a laboratory at the level that is required for the generation of electric power at a nuclear plant. Accordingly, the nuclear fuel cycle in a laboratory environment was completed.

At that time, Iran, during a special ceremony, injected uranium-hexafluoride into three thousand centrifuges, and celebrated her entry into the list of the countries that produce nuclear fuel. In 2006, the news of the installation of six thousand new centrifuges was announced. Two years later, news was also made public of the packaging of the fuel and its preparation and placement in the reactors for power generation, in addition to the testing of two new centrifuges with a several-fold increase in capacity compared to the older ones. Toward the end of 2008, the last Iranian nuclear fuel cycle was completed in the nuclear plant in Isfahan. In the winter of 2011, the use of the Fardo nuclear site for enrichment, the success in enriching uranium to 20 percent, and the search for 10 new sites for the construction of enrichment facilities were announced.

In April 2011, the third generation of centrifuge machines, and a sample of virtual fuel for the research reactor in Tehran were showcased. The productivity of these new centrifuges is 5 times higher than the first-generation centrifuges. Furthermore, on February 15, 2012, the Iranian official news agency, Fars, announced that new domestically produced fuel rods had been lowered into a Tehran light water research reactor pool.

Construction of Fuel Complex

In 2008, the Fuel Manufacturing Plant (FMP) began operations to produce pellets, rods, and a fuel complex, and came up with the first fuel complex for the 40-megawatt reactor in Arak. This factory is capable of an annual production of 10 tons of nuclear fuel for use in the heavy water research reactor in Arak, and 30 tons of nuclear fuel with a maximum of 5 percent enrichment for the light water reactors, such as the 360-megawatt reactor in Darkhovein or the Bushehr nuclear power plant. This factory has the capacity to increase fuel production to supply 2,600 megawatts of electricity under the Fifth Development Plan.

Production of Deuteron Compound

In December 2010, Iranian experts produced deuteron for the first time, which is a stable particle composed of one neutron and a proton and which has many applications in scientific research. The heavy water that is produced in the Arak complex is the source of deuterium, which is used in the synthesis of deuteron compounds, and Iranian

Table 8.1 Nuclear Technology Achievements in Iran

Row	Achievement	Name of the Institute
1	Production of Iodine 131 radiopharmaceutical for diagnosis and treatment of thyroid diseases Phosphorous 32 in the form of Colloidal and oral solution, ^{14}C- urea, EDTMP -153Sm (Ethylene Diamine Tetra (Methylene Phosphoric Acid) Sodium)	Nuclear Sciences Research Institute
2	Production of radiopharmaceutical kits	
3	Labeling of antibodies and peptides with radionuclides	
4	Production and quality control of low dose rate (LDR) Brackytherapy sources Iridium 192 for the treatment of soft tissues cancer	
5	Acquiring hydrogel technology for medical purposes	Radiation Application Research Institute
6	Determining of sterilization dosage and providing sterilized products quality control services	
7	Variety of radiopharmaceuticals and radioisotopes such as: Tallium 201, Gallium 67, Krypton 81, Indium 111, Cobalt 57, Palladium 103, Copper	Agricultural, Medical and Industrial Research Institute
8	Calibration of radiotherapy instruments	
9	Production of molybdenum generator	Materials Research Institute
10	Production of various outer protective clothing for use in nuclear industry	Plasma Physics Research Centre and Islamic Azad University, Science and Research branch
11	Production of various advanced ceramics for applications in nuclear industry	Researchers of Tehran technology park
12	Production of "Iodine 131" radiopharmaceutical for diagnosis and treatment of tumors	Researchers of Medical Science and Nuclear Science and Technology Research Center
13	Production of "Rhenium 186" radiopharmaceutical for relief of pain due to cancer	
14	Production of "Peptide Ubiquicidin" radiopharmaceutical for diagnosis of infection	
15	Production of "Peptide Bombesin" radiopharmaceutical for early diagnosis of breast, prostate, and lung cancer	
16	Production of laser machine for breaking of kidney stones	Center for Laser Science and Technology
17	Construction of prostate laser machine	
18	Instruments of production of laser welding	
19	Construction of skin laser machine	Laser Institute

Source: Bal, 2011.

researchers' achieving self-sufficiency in the production of heavy water with the synthesis of deuteron solvents satisfied the Iranian scientific community's needs for measuring instruments. The deuteron solvents have applications in Nuclear Magnetic Resonance (NMR) instruments, as well as in chemical and medical studies. Research relating to deuteron was started in cooperation with the universities in the Arak Heavy Water complex in 2008, and in 2010, five varieties of the deuteron compound that are needed by the Iranian scientific and research centers were produced domestically.

We present some of the achievements of AEOI nuclear research institutes in Table 8.1 (Bal 2011).

We note that the Institute for Nuclear Fusion is one of the 8 institutes of Nuclear Science and Technology Research Center. More than one hundred fifty researchers are employed at this Institute, and Tokamak (a special nuclear reactor) is used for research there.

At the time of the writing of this chapter in March 2012, approximately four hundred large and small research projects are underway at the Nuclear Science and Technology Research Center. The radiopharmaceutical products that are produced by the Center are used for the treatment of between eight hundred thousand and one million cancer patients at over 125 hospitals around the country.

A List of Advanced Nuclear Technologies and Capabilities in Iran

Iranian scientists have been actively involved in the technological development and commercialization of their nuclear achievements. Below we present these achievements:

- Training of many nuclear engineers and scientists
- Identifying suitable locations and preparing special sites and facilities for the storage of nuclear wastes
- Exploring and extracting uranium ore
- Producing yellow cake at a factory in Bandar Abbas, and constructing another factory forthe production of this substance in the city of Ardakan in central Iran
- Acquiring the technology of yellow cake production and the separation of the impurity of Molybdenum from yellow cake
- Producing uranium products that are needed for the fuel cycle, such as Uranium Hexafluoride (UF6) and Uranium Dioxide (UO2) with different levels of enrichment

- Constructing enrichment facilities in Natanz, and enriching hexafluoride to 20 percent enrichment level
- Constructing another factory in Fardo
- Constructing a factory for the production of the last chain of the fuel cycle
- Constructing the research reactor IR40, which is almost complete
- Placing a new generation of centrifuges into use
- Producing and using Zirconium in the production of fuel rods
- Producing ultra pure magnesium with 99.99 percent purity[3]

Producing heavy water and its special facilities, which are underway[4]

- Manufacturing dosimeter tools and equipment as well as developing the optimal design of monitoring devices for measuring ionization radiation and manufacturing these devices
- Acquiring a great deal of knowledge for the design and construction of nuclear accelerators
- Developing capabilities in the design and construction of Gamma irradiation systems.
- Establishing some of the infrastructure for nuclear research, such as a cyclotron, and developing skilled manpower resources
- Establishing a center for nuclear security and protection against radiation
- Producing a variety of instruments for measuring radiation, including tools for measuring contamination or whole body counters.

The Challenges Ahead

In spite of these major achievements, the Iranian nuclear industry also faces many challenges. We list these challenges below (Institute of Energy 2007):

- Fewer experiences in the development of technology of nuclear power generation in Iran compared with those developing countries that currently have access to nuclear power plants.
- Economic sanctions and the dependence of nuclear power plants on certain foreign-manufactured products, and inadequate manufacturing capabilities in the production of power plant equipment.
- Limited ability to explore for uranium ore, which currently covers only one-third of the area of the country.
- Inadequate depth of knowledge, technology, culture of security, and weakness of quality control in some nuclear technology sectors.

- Unfavorable environment for international cooperation in the development of nuclear power plants.
- Limited success in the commercialization of nuclear innovations and inaccessibility to the regional nuclear markets because of the sanctions.

Future of Nuclear Technology in Iran

In the face of some of the most severe sanctions in the history of mankind, Iran has acquired technological capabilities in nuclear technology, ranging from the construction of a nuclear power plant to the complete progression of the fuel cycle, and from industrial applications of nuclear energy to the medical use of the technology. Most of these gains have been achieved through a reliance on domestic resources. The technology planners have set the strategic national goal of making Iran the leading nuclear power in the Middle East and Western Asia.

Without doubt, Iran can further develop the technology in the near future, and will become one of the key nuclear technology players in the Middle East, Africa, and Western Asia.

Acquiring advanced nuclear technology has been elevated to the level of a national aspiration of the Iranian people. The economic sanctions against Iran, because of the country's relentless pursuit of nuclear technology in the face of unparallel adversity created by the leading Western powers, have turned the goal of the acquisition of the technology into a symbol of national resistance against the excess of foreign intervention in domestic Iranian affairs, a goal that will not be altered by changes in the administration. This national desire is similar to the Iranian desire to nationalize the Iranian oil industry 60 years ago.

In addition to the resolute will of the Iranian government and people to acquire full use of nuclear technology for peaceful purposes as a sign of national of national self-determination and independence, the country has the aim of commercializing its nuclear technological achievements over the last decade. An important path for further development of the industry is the formation of a consortium, with the cooperation of the advanced nuclear powers, for entry into the regional nuclear markets.

Long-Term and Midterm Goals of AEOI

We enumerate the long- and medium-term goals for nuclear technology development in Iran below:

Long-Term Goals:

- Performing studies and taking the necessary measures to locate, design, construct, and safely operate nuclear power plants to supply 10 percent of the electricity needed in Iran
- Supplying part of the fuel needed for the operation of twenty-thousand-megawatt nuclear power plants and research reactors.
- Developing and commercializing other applications of the nuclear industry (radiopharmaceuticals, irradiation systems and research reactors)

Midterm Goals (in the Next Five Years):

- Completing studies on uranium, which covers one-third of the area of Iran
- Beginning the exploitation of two uranium mines
- Constructing a new factory that produces yellowcake in the mines, which will be explored in the future
- Producing fifty thousand centrifuges
- Constructing a waste disposal facility
- Producing new radiopharmaceutical applications
- Designing, constructing, and operating five irradiation systems
- Constructing and operating accelerators for production of radiopharmaceuticals
- Designing and constructing 10- and 20-megawatt research reactors
- Commissioning and operating a 40-megawatt research reactor
- Acquiring the technology of the design, construction, and the commissioning of a 360-megawatt nuclear power plant (to generate electricity)
- Constructing at least two 1,000-megawatt power plants in collaboration with foreign partners
- Producing nuclear fuel for existing research reactors and the Bushehr power plant
- Developing main nuclear codes

Summary and Conclusions

In this chapter, the history of the development and the plans and technical advances of nuclear technology in Iran was reviewed. We discussed how the United States eagerly recommended to the shah of

Iran that the country should purchase many nuclear power plants for the generation of over twenty thousand megawatts of electricity. Other Western countries, particularly France and Germany, were also eager to sell turnkey nuclear power plants to Iran, and concluded agreements to cooperate with Iran in the development of nuclear energy in the 1970s. However, all of these agreements were only in the form of sales of nuclear facilities, and were not designed to transfer nuclear technology to the country.

After the triumph of the Islamic Revolution in 1979, two developments led to the abandonment of efforts to generate nuclear power in Iran. First, the revolutionary government in Iran abandoned the shah's policy of purchasing turnkey nuclear power plants. Second, fearing the radicalism of the new revolutionary regime in Iran, the Western countries, especially the United States, adopted the policy of isolating Iran and depriving the country of this particular source of energy and other technological development. These policy changes, as well as the eight years of national defense against the Iraqi invasion of Iran, did not permit Iran to pursue the acquisition of nuclear technology. It was only in the early years of the 1990s that Iran resumed efforts to acquire this technology.

The change in policy in the early 1990s to enable Iran to acquire the technology was based on the realization that Iran cannot neglect the massive direct and implicit costs of not using nuclear energy. On the one hand, because of the rapid growth of the population and the equally high rate of industrialization and economic growth, Iran faces a very high demand for energy. On the other hand, meeting the ever-increasing demand for energy by relying on very scarce natural, nonrenewable resources such as natural gas and petroleum creates massive opportunity costs for the country. It is estimated that at the current rate of extraction, Iran cannot continue to be an exporter of energy, and may become a net importer of energy within the next four decades. Additionally, the use of fossil fuels to generate energy is economically inefficient in the sense that highly scarce natural gas and petroleum can be used to produce petrochemical products, which would generate substantially higher value-added and create higher employment and income in the country compared to the burning of fuels for electric generation. More importantly, the technological development of the country on all fronts requires cohesiveness, like a closely linked chain, and in the future, knowledge of nuclear fission and fusion will constitute a central focus of the scientific and technological development of countries. Iran is aiming to close the technological gaps between itself and the advanced industrial countries that it currently faces, and it certainly does not desire to fall behind the rest of the world in this important field by abandoning its efforts to master nuclear technology.

As discussed in the chapter, in spite of severe international sanctions, Iranian researchers and technologists have made notable progress in many areas of nuclear technology. Development of these technologies was made possible through the cooperation of many entities and with the strong support and guidance of the Supreme Leader, Ayatollah Khamenei, the executive branch under the leadership of President Ahmadinejad, and the legislative branch of the government. Among these entities we can name the Atomic Energy Organization of Iran, the universities, the nuclear research centers, and the companies that produce parts and instruments for use in nuclear R&D, all of which are important players in promoting this industry.

The IR of Iran does not demand extra-legal rights for the development of nuclear technology. Being fully committed to the principles of the United Nations NPT, Iran insists on protecting its rights as a member state to pursue the development of nuclear technology for peaceful purposes. Iran's position on fully realizing its rights under the Treaty is transparent and clear, and the country has steadfastly attempted to create an environment of trust in the international community with respect to its intent for the peaceful use of nuclear technology.

The sizable number and high technical competency of the human resources, particularly in nuclear physics and nuclear engineering, in Iran, and the enormous investment in ongoing research projects promises a bright future for the development of this key technology in Iran. Moreover, these investments in human and physical capital not only point to the reality of Iran's presence in the field of nuclear technology but also to country's taking a prominent role in the regional nuclear market in the future.

Notes

1. A version of a chemical element that has an unstable nucleus, which in the processes of decaying to a stable form emits radiation.
2. Due to the sanctions, the foreign firms in Iran operate surreptitiously.
3. This chemical has many uses in the automotive, defense, and nuclear industries, and is used in the production of aluminum alloys, extraction of special chemicals, and cathode protection.
4. Heavy water is needed for cooling and slowing the research reactor IR40.

References

Aghamiri, S. M. 2010. *Nuclear Energy: The National Intent*. Tebyan Department of Science and Technology. Tebyan.net. October 27, 2010. www.tebyan.net/social/experts/scientificexperts/2010/10/27/141711.htm.
Alikhani. M. 2006. "Looking at the History of Iran's Nuclear Programs." Iranpress.ir. July 4. Accessed August 23, 2012. www.iranpress.ir/egolestan/egolestan/News.aspx?NID=570.
Bal, Z. 2011. "19 Nuclear Achievements of the Country/ 22 Radiopharmaceutical Were Produced" (in Farsi). Mehrnews.com. April 8. www.mehrnews.com/fa/NewsDetail.aspx?NewsID=1283651.
Documents & Diplomatic History Center. 2007. *Nuclear Energy History in Iran & World*. Tehran: Ministry of Foreign Affairs Press Center.
Institute of Energy. 2007. *National Development Plan of Nuclear Power Plants In Iran* (in Farsi). AEOI.org. Accessed August 23, 2012. http://www.aeoi.org.ir/portal/Home/Default.aspx?CategoryID=5924979f-6434-45cb-b5ac-ed27b218368e.
International Atomic Energy Agency (IAEA). 2011. *IAEA Annual Report 2010*. IAEA.org. Accessed August 29, 2011.
Iran Expediency Council. 2003. *General Policies of Iran's Fourth Development Plan*, Tehran: Iran Expediency Council.
Iran's Department of Research and Commentary. 2011. *Iran Achievements in Various Areas of Nuclear Technology*. Irna.ir. Accessed August 23, 2012. http://www.irna.ir.
Iran's Fifth Five-Year Development Plan. 2010. The Office of the Vice President of Strategic Planning and Monitoring. Accessed August 23, 2012. http://rc.majlis.ir/fa/law/show/790196.
Iran's Fourth Five-Year Development Plan. 2004. The Office of Vice President of Strategic Planning and Monitoring. Acessed August 23, 2012. http://rc.majlis.ir/fa/law/show/94202.
Karkhanehei, E. 2010. "Nuclear Energy and Scientific Development of Iran." Edunews.ir. Accessed August 23, 2012. www.edunews.ir/index.php?view&sid=24441.
Koch A., and J. Wolf. 1997. "Iran's Nuclear Procurement Program." *The Nonproliferation Review* 5 (1): 123–135. DOI:10.1080/10736709708436700.
Mohammadi, H. 2011. "Browsing the Nuclear Activities of Iran from Beginning Till Now." Afkarenow.com. Accessed August 23, 2012. www.afkarenow.com/mataleb/hasteifaaliat.htm.
Mohammad Hosseini Targhi, M. 2001. *Rules And Regulations of Nuclear Energy*. Tehran: Atomic Energy Organization of Iran.
National Education Assessment Organization. 2011. *Manual for Selection of Undergraduate Degree Courses*. Sanjesh.org. Accessed January 6, 2011. http://www.sanjesh.org.
———. 2011. *Manual for Selection of Master's Degree Courses*. Sanjesh.org. Accessed August 23, 2012. http://www.sanjesh.org.

———. 2011. *Manual for Selection of Doctoral Degree Courses.* Sanjesh.org. Accessed August 23, 2012. http://www.sanjesh.org.

The Office of Higher Education. 2011. "Statistical Details of Nuclear Physics at the University: The Number of Students and Graduates." Mehrnews.com. Accessed August 23, 2012. www.mehrnews.com/fa/newsdetail.aspx?NewsID=1284549.

Sahimi, M. 2005. "Iran's Nuclear Program and Enhancing the Technical Basis." Hamshahrionline. Accessed August 23, 2012. http://www.hamshahrionline.ir/hamnews/1384/840501/world/sciew.htm.

Secretariat of Scientific Cooperation Council of Atomic Energy Organization of Iran. 2008. *Iran Nuclear Industry at a Glance.* Tehran: Secretariat of Scientific Cooperation Council of Atomic Energy Organization of Iran.

Soltani, B. 2011. "Nuclear Technology Management in Iran." Paper presented at Fifth National and First International Conference of MOT in Iran, November 22–23. Tehran.

CHAPTER 9

Iran's Aerospace Technology

Parviz Tarikhi, Mohammad Abbassi, and Maryam Ashrafi

Introduction

Aerospace is defined as the science of Earth's atmosphere and space beyond, and the technology of flight and space travel. It deals with the design of airplanes and spacecraft, as well as related topics and issues. Topics in the field of aerospace engineering are among the most advanced research subjects, and they attract large sums of research and development funds from civilian and military sources. Research in this field also boosts progress in other engineering areas. Aerospace is a multidisciplinary branch of science and technology, which draws on knowledge from physics, mechanical engineering, metallurgy, computer science, electronics, and other subjects.

Realizing the significance of the aerospace industry, the Iranian government placed further development of this industry on its agenda starting in the Fourth Five-Year Development Plan (2004–2010). The aerospace industry was given top priority in the technological development of the country in the General National Scientific Plan, which was formulated in 2010. In this Plan, the goals, objectives, and quantitative indicators of the success of strategies for developing this technology were taken into consideration.

In the next section of this chapter, the background of the scientific and technological activities in the field of aerospace in Iran will be discussed. The second section will elaborate and analyze the current status of these technologies, the main actors in the Iranian aerospace industry, the infrastructure, human resources, international cooperation, and research and development (R&D) funding. In the third section, future trends and plans for development of the industry are examined, and the final section will make concluding remarks on the Iranian aerospace industry as an industry of high priority for Iran's science and technology policy makers.

Background

The desire to fly up to the sky and voyage in space is a common idea not only in national myths but also in religions. The significant contribution of the Iranians to this mythology is undeniable. In an Iranian myth, for example, Kay Kāvus flies up to heaven using propulsion by hungry eagles, as described by Ferdowsi, one of the world's greatest epic poets, in his magnificent work *Shah-Nameh* (Tarikhi 2009). However, in contemporary times, Iran's attention to this field and commitment to developing the infrastructure for the aviation industry goes back to the 1930s, when the Shah Reza Pahlavi founded it with the assistance and expertise of the German Junkers & Co Aviation. The field was further expanded in the 1970s during the reign of Pahlavi. [Attention: please note that Shah Reza Pahlavi and his son Muhammad Reza Pahlavi were different and therefore it is needed to keep Shah Muhammad Reza here and do not delet it. Tarikhi] He not only ordered a large number of advanced weapons manufactured in the United States but also attempted to acquire the capability to assemble them domestically. Bell Helicopter, a division of Textron, Inc., thus built a factory in Isfahan to produce Model-214 helicopters, while Northrop was a joint partner in Iran Aircraft Industries, Inc., which maintained many of the US military aircraft sold to Iran, and was expected to produce aircraft components and eventually complete planes. However, as was the case in automobile manufacturing (please see chapter 10 on the automotive industry in this volume), these aircraft "manufacturing" activities were nothing more than the assembly of foreign-produced aircraft parts in Iran. In short, the Iranian firms involved in the assembly of the aircraft were devoid of any technological capabilities and were considered to be "latecomer" firms[1] (Wikipedia 2012a).

Commercial development of the aviation industry in Iran started as early as 1923 with the establishment of the first airliner office in Tehran, in cooperation with Junkers & Co Aviation, The company provided air travel services between Tehran, Mashed, Shiraz, Bandar Anzali, and Boushehr. [Attention: I suggest that the name Shah Reza is not deleted to avoid confusion. Tarikhi] Pahlavi established an office to launch the aviation industry in 1922, and it was the first official aviation organization in Iran.

Starting in 1958, Iran showed an interest in using technologies for peaceful applications and the exploration of space for economic and social development when the country joined 17 other member states to establish the United Nations ad-hoc Committee for International Cooperation on Space. The aims of this organization, which would later change its name to the Committee on the Peaceful Uses of Outer Space (COPUOS), were to review international collaborative programs aimed at exploiting and using space for civilian purposes, serve as a forum for information exchanges, and encourage the development of national programs to study outer space (UNOOSA 2009).

In 1969, by establishing the Asad-Abad Ground Station and installing a 30 m-diameter standard-A antenna to connect to the Pacific Intelsat for an international communications network, Iran became a participant in the American telecommunications global network. In 1972, following the launch of US ERTS (Landsat-1), the United States provided technical assistance to Iran to construct the Mahdasht Satellite Receiving Station (MSRS), which at the time was one of only five data-receiving stations around the globe (Telecommunication Company of Iran 2008). This was Iran's first act of bilateral cooperation in space technology.

The common applications of space technology, consisting of telecommunications, television broadcasting, remote sensing, navigation, tele-education, weather forecasting, environmental modeling, the Internet, and relief and rescue operations were widely adopted in Iran. In 1974, Iran and the General Electric (GE) Company of the United States entered into an agreement for the installation and operation of a satellite data-receiving station. However, at the start of the revolutionary upheaval in Iran in 1978, GE ended its commitment to provide technical assistance for the installation and operation of the facilities for tracking Earth resource satellites and direct data acquisition from those satellites (Tarikhi 2008a).

The initial Iranian efforts to establish independent space projects began in 1977 when the country aimed to develop an Iranian communications satellite system called Zohreh. In spite of the presence of a number of national organizations in the development of plans to send research satellites into space, Iran could not achieve these goals entirely indigenously and needed foreign assistance in the required technological fields. Facing Western countries' refusal to supply the technologies, Iran turned to the Soviet Union, China, and India, which at that time were the leading nonwestern space-faring countries. Eventually, North Korea and later Italy were Iran's partners in space research and development.

Along with the plan to establish Zohreh, the first communications satellite system, Iran aimed to create the Iranian Space Agency (ISA) in 1977 (Tarikhi 2003). However, the unstable revolutionary conditions in the country, and the protracted eight-year war with Iraq, beginning in 1980, all but ended the efforts to institutionalize space activities in Iran. What remained were some pursuits relating to space applications, such as communications and remote sensing (Tarikhi 1994).

Current Status of Iran's Aerospace Technology

After the Western sanctions were imposed following the Revolution in Iran, the country shifted its general policy for aeronautics from the goal of purchasing the world's best available equipment and technology in the field to being able to manufacture it independently in order to meet domestic needs, especially of technological products, and therefore becoming "sanction-proof." Practically in no other field was this urgency greater than in aeronautics. Therefore, Iran has avoided the urge to purchase better Western aircraft available to it from time to time in favor of aircraft that could be manufactured at home through arrangements of purchasing licenses and technologies, as well as reverse-engineering parts (Wikipedia 2012b).

Nevertheless, in the field of space, shortly after the launch of the first US Earth observation system, Landsat, Iran, with technical assistance from the United States, established the Iranian Remote Sensing Center (IRSC), which has as its mission the collecting, processing, and distributing of relevant imagery products to users throughout the country for resource planning and management (Tarikhi 2006; Tarikhi 2007). Having access to remote sensing data assisted Iran in dealing with an array of technical problems including identifying areas suitable for economic development and pinpointing areas prone to earthquakes, floods, landslides, and other natural disasters; investigating greenhouse gas emissions and air pollution in the large urban areas; monitoring wetlands and water basins inland and those shared with neighboring countries; and other useful activities for global benefits.

Actors in Policy Making for Aerospace Technology

Currently the Iran Aviation Industries Organization (IAIO) acts as a policy maker and coordinator to promote an indigenous Iranian aeronautical industry by providing and assisting the Iranian aircraft industries with needed technologies, knowledge, and parts. IAIO was established in 1966 for the purpose of planning, controlling, and managing Iran's noncivilian aviation industry. Under its umbrella, IAIO has five aviation organizations including SAHA, HESA, PANHA, GHODS (the acronyms are from the Farsi names), and the Shahid Basir Industry, and each one plays a different and complementary role in the country's aviation development (Wikipedia 2012b).

The structure and role of policy-making bodies for the development of aerospace technology have changed somewhat in recent years. These policy makers include Ministry of Post, Telegraph and Telephone (MPTT), the Iranian Broadcasting Organization, and the Ministry of Science, Research and Technology, which cooperate for the purposes of telecommunications and broadcasting, as well as other applications. However, the full institutionalization of such efforts did not occur until February 2004, when the ISA began its activity according to Article 9 of the Law for Tasks and Authorizations of the Ministry of Communications and Information Technology (MCIT), which was passed by the Iranian Parliament on December 10, 2003. The president of the ISA held the position of Vice-Minister of MCIT and of the secretariat of the Space Supreme Council (SSC) simultaneously (Tarikhi 2008b). The ISA's mission consisted of monitoring and supporting activities related to peaceful applications of space science and technology under the leadership of the SSC, chaired by Iran's president (Tarikhi 2008c).

The establishment of the ISA is considered a major practical step toward making advances in relevant science and technology that will lead to the effective use of outer space for peaceful purposes. It also plays an important role in promoting international cooperation for this highly desirable purpose. Some of the most important tasks assigned to the ISA by the SSC are listed below:

1. Implementation of studies, research, design, and engineering in space services, remote sensing, strengthening domestic and international space networks;
2. Preparation of medium- and long-term space exploration plans;
3. Conducting of studies and research on design, construction, and launching of satellites.

With the goal of increasing the efficiency of the space administration, the Iranian state decided to make a number of organizational changes, including the dissolution of the SSC and subsequently, the approval of a new statute for the space agency. These changes took place during 2007 and 2008, and resulted in the creation of the Science, Research and Technology Commission that operates within the Cabinet of the President of the Islamic Republic (IR) of Iran. Under the new organizational structure, the ISA operated under the Ministry of CIT and did not report to the President of the Republic directly. Moreover, the new statute provided the ISA with more funds so that it could focus efforts in reaching its goals. Furthermore, the new statute authorized the ISA to establish space research centers and spin-off firms, conditional upon receiving authorization from the Council for Development of Higher Education. The Presidential Cabinet authorized the ISA to receive approved tariffs for the offering of space services (Tarikhi 2008d).

After the dissolution of the SSC, and because of political issues, the discussion of which is beyond the scope of this chapter, the Iranian Parliament considered the dissolution of the councils unlawful. The matter was forwarded to the supreme arbitrating body, the Expediency Council, and the Council decided to revive it on September 27, 2008. Accordingly, the Council mandated the executive branch of the government to revive the dissolved SSC eight months after its dissolution. Finally, the Iranian Vice-President for Science and Technology established nine agencies to facilitate and integrate the development of science and technology into the fields within their administrative purview, one of which was the administration of aerospace technology development that commenced in 2009. The members of each agency consist of 20 senior managers and experts from government, the private sector, academia, and scientific organizations that are involved in the aerospace sector (Tarikhi 2011b).

Actors in the Technological Field

To attain their goals in the field of aerospace, many research centers and organizations are active. In aeronautics, under the IAIO, the Iran Helicopter Support and Renewal Company (IHSRC), known by its Persian acronym PANHA, was established in 1969. The Iranian Aircraft Industries (IACI), known as SAHA, was founded in 1970, and the Iran Aircraft Manufacturing Industries Corporation (IAMI), known by its Persian acronym HESA, in 1974. Two other important companies, the Iranian Armed Forces Aviation Industries Organization (IAFAIO), and Ghods Research Center were formed in the early 1980s. By September 2004, the Iran aviation industry had produced more than 1,600 aircraft; 2,182 aeroengines; 1,751 helicopter engines; and 149 industrial jet engines, in addition to repairing more than 11 models of aircraft and 18 models of military, commercial, and industrial aircraft engines.

Iran has domestically designed and manufactured Azarakhsh and Saeqeh fighter jets through mass production, and it has plans to extend these efforts to the launching of helicopters, turboprops, and passenger planes. The country has produced a Boeing 737–800 simulator that is the first in the Middle East. In order for the country to continue on this path, an effort is underway to secure additional passenger planes to meet the country's needs for 6,300 airplanes for the Iranian population of 75 million. Currently, Iran possesses only nine aircraft per one million individuals (Wikipedia 2012b). The background and organizational structures of the most important of these centers in aeronautics and space science and technology are discussed below.

The IAIO and Its Subsidiaries
As one of the agencies of the Ministry of Defense and Armed Forces Logistics, the IAIO, as mentioned above, consists of five subsidiary organizations, including the Organization of Aircraft Industries of Iran, the Helicopter Supports and Renovation Company of Iran, the Aircraft Manufacturing Company of Iran, Quds Industries, and the Research Institute of the Organization of Aircraft Industries. IAIO and its subsidiaries are involved in building jet engines; building parts for a variety of aircraft; repairing, maintaining, and overhauling passenger planes; and constructing hangars for wide-body aircraft. It is considered to be the largest company in this field in the Middle East (Aerospace Technology Development Administration 2009). Some of the activities of the above listed organizations are discussed below:

Helicopter Supports and Renovation Company of Iran
This company is the largest of its kind in the Middle East. In the military aviation sector, the company maintains and repairs helicopters of makes such as Bell 205, 206, 209, 212, 214, and 412, as well as CH-7, RH-53D, SH-3, and MIL-171, according to the military maintenance standards. In the civilian aircraft maintenance field, this company is the only firm that has received a license from the National Aircraft Organization for maintenance efficiency. This company also produces helicopters, black boxes, floating systems, and many other aircraft parts (Aerospace Technology Development Administration 2009).

Iran Aircraft Manufacturing Company
Through technology transfer from Ukraine, this company manufactures Iran-140, a 52-seat passenger airplane with jet propeller engines and a two-thousand-km flight range. Moreover, this company designs

and manufactures a variety of drones, fuselages, and other aircrafts (Aerospace Technology Development Administration 2009).

Quds Industries
This company, in addition to designing and manufacturing a number of aircraft and types of equipment such as drones, propeller aircraft, parachutes, and automatics, as well as manual parachutes for drones; ground control systems; (GCS) avionics systems; and aerial photography, targeting, and optic guidance systems, also provides after-sales services. Moreover, this company designs and manufactures light two-, four- and eight-seat aircraft (Aerospace Technology Development Administration 2009).

Research Institute of the Organization of Aircraft Industries
The research programs of this organization consist of the design of piloted and pilotless aircraft, the simulation of aerodynamic processes in computational fluid dynamics laboratories, the conducting of aerodynamics tests, the development of aviation products using the national wind tunnel, the design of systems for launching and retrieving aircraft, the standardization and validation of avionic products for training purposes, the promotion and development of laboratories related to the aviation industry, and the evaluation, control, and auditing of aviation projects (Aerospace Technology Development Administration 2009).

Iran Aviation and Space Industries Association (IASIA)
This association has 130 members that are the companies active in the aerospace industry. The member companies have six thousand employees with expertise in various aerospace fields, and they operate with an annual budget of 500 million Euros and are active in Iran's aviation and space industry (Ebrahimi 2011).

Iranian Space Agency (ISA)
The ISA is the only national (governmental) space agency of Iran. It was annexed to the Presidential Institution on September 29, 2010, by the approval of the Iranian Administrational Supreme Council. The ISA is responsible for the execution of space policy throughout the country. It conducts research in the fields of space technology, such as remote sensing and communications.

The Aerospace Research Institute of Iran (ARI)
To conduct research in aerospace, the Ministry of Science, Research and Technology established the ARI of Iran in 2000. To fulfill the

country's research needs in this field and establish connections with the related industries, ARI has conducted a notable set of activities. The ARI has the following goals (ARI n.d.):

- Recognition and introduction of aerospace technologies, and cooperation with the related organizations and entities for acquisition of the latest aerospace technologies;
- Development and expansion of research in the aerospace field and attempts to meet the research needs of the country;
- Research cooperation with research and educational organizations of the country with the aim of improving the quality of related research activities;

The ARI has created the necessary environment for research and established increased research facilities for researchers, which consist of a parallel processing laboratory, electronic laboratory, virtual reality laboratory, Information Technology Center, construction and assembly plants, and library. The ARI has worked under the ISA since the annexation of ISA to the Presidency Institution of the IR of Iran (Tarikhi 2010).

Iranian Research Organization for Science and Technology (IROST)
The main goal of the IROST is to support the development of technology through research and development at the national level. To achieve this goal, it offers scientific, technical, financial, legal, administrative, and cultural support to applicants. Furthermore, the IROST creates conditions for the efficient and effective interaction of the demand for and supply of technology so that fertile ground for the growth of creativity and innovation, the use of the results of research, and the commercialization of the technologies resulting from research and development are generated in a competitive environment. The aerospace mechanics group of the IROST, housed in the Mechanics Institute, is one of the six institutes of the organization, which is responsible for important projects, including the design and construction of Mesbah Satellite (IROST n.d.).

Iranian Aerospace Society (IAS)
The goals of the IAS are to engage in activities relating to scientific development, research, and specialized technical aspects of aerospace for peaceful purposes. The society was established in 1993 (IAS n.d.).

In addition to the above listed entities, the Iranian industries and companies related to space development and production are as follows:

Aerospace Industries Organization (AIO) (Sazemane Sanaye Hava-Faza [SSH])
The AIO is the leading high-tech industrial and military subsidiary of the Sanam Industrial Group. It is the manufacturer, among others, of Shahab ballistic missiles, launchers, rocket/booster propellants and components; it also supplies nonmilitary items and services such as fuel pumps, technical and engineering services, and research and development (R&D). The AIO is the obvious organization to lead the development and production of the space assets of Iran. It manages a number of factories and research centers throughout the country (Harvey et al. 2010).

Shahid Hemmat Industrial Group (SHIG)
Shahid Hemmat Industrial Group is subordinated to the AIO, and is comprised of several divisions that are related to building and operating launch vehicles, such as the Kalhor Industry, Karimi Industry, Cheraghi Industry, Rastegar Industry, Varamini Industry, and Movahed Industry.

Iran Electronics Industries (IEI) (Sanaye Electronic-e Iran [SEI]; SAIran)
Iran Electronics Industries was established in 1972, and is presently the major producer of electronic systems and products in Iran. Subsidiaries are Shiraz Electronics Industries; Iran Communication Industries; Information Systems of Iran; Electronic Components Industries in Tehran and Shiraz; Isfahan Optics Industries; Iran Electronics Research Center; Iran Space Industries Group; and Security of Telecommunication and Information Technology (STI).

Iran Space Industries Group (ISIG)
The founding of the ISIG was announced on the occasion of the launch on February 4, 2008, of the Kavoshgar-1 rocket. ISIG is subordinated to the IEI.

Shiraz Electronics Industries (SEI)
Since 1973, SEI has been professionally engaged in developing electronic products and projects. It employs highly skilled personnel, and with their experience with advanced equipment, and their abundant know-how and motivation, the company has formed a well-organized and powerful technological industrial group. SEI is currently involved

in electronic warfare, control and automation, radar and microwaves, weapon electronics, avionics, computers, and electro-optics.

Iran Communication Industries (ICI) (Sanaye Mokhaberat Iran [SMI])
The ICI is Iran's leading manufacturer of military and civil communications equipment and systems. The company produces more than 75 products in the field of tactical communications and encryption systems that meet a wide range of military requirements.

Information Systems of Iran (ISIRAN)
ISIRAN is a state-owned company that was founded in 1971. It is one of the largest and most experienced information companies in Iran. This company claims to be Iran's number-one IT company in terms of revenue, market share, and variety and quality of products and services. ISIRAN provides assistance and consulting services to its clients on information systems.

Electronic Components Industries (ECI)
The ECI was founded in 1976, and is comprised of two complexes, one in Shiraz and the other in Tehran. Its activities include, among others, the design and manufacture of semiconductor devices, quartz crystals, multilayer PCB and thick film hybrid products, infantry field wire, optical cable and access systems.

Isfahan Optics Industries (IOI)
The IOI was founded in 1987 with the aim of setting up a vigorous and modern optics industry. The employment of highly qualified engineers and state-of-the-art equipment has reportedly made the IOI one of the most capable industries in Iran. The design and manufacture of complex lenses and prisms, multilayer coatings, a wide range of day light sights, and various types of aircraft windshields are carried out at the IOI.

Iran Electronics Research Center (IERC)
The IERC, founded in 1997, is a scientific, educational, and research institute, with active research teams in the fields of electronics, communications, microprocessors, microelectronics, optics, electro-optics and radars. The research groups at the IERC include electronics, communications, microprocessors, microelectronics, optics and electro-optics. The center is capable of handling the multitechnology range of large products through an advanced PERT system.

Shahid Bagheri Industrial Group (SBIG)
The SBIG is part of the Defense Industries Organization (DIO), and it cooperates with Russia's Baltic State Technical University and the Sanam Industries Group to run the Persepolis (Takht-I-Jamshid) joint missile education center in Iran with the purpose of transferring missile technology from Russia to Iran.

Iran Telecommunication Manufacturing Company (ITMC)
The ITMC, with the goal of producing high-capacity telecommunications centers and telephone systems, was established in 1967 and became operational in 1969.

Telecommunication Company of Iran (TCI)
The TCI is the country's largest company involved in telecommunication issues, and it functions under the Ministry of Communications and Information Technology (CIT). It has branches in almost all of the provinces throughout the country, and is responsible for the development and management of the country's communications infrastructure using communications technology, particularly satellite- and ground-based telecommunications.

National Cartographic Center (NCC)
Established in 1953, the NCC, with more than half a century of experience, is the main authority for the production of maps and spatial information under the IR President's Deputy for Planning and Strategic Supervision. Benefiting from the expertise of eight hundred well-trained personnel, the NCC supervises and oversees technical control of mapping and spatial information projects that are carried out by the NCC, other governmental organizations, and private mapping companies. Production of the base map of Iran; marine charts; the design and establishment of the National Geodetic Control as well as geodynamical networks; the establishment of national, regional, and urban spatial topographic databases; and the production of small-scale base maps and national atlases are examples of what the NCC has accomplished. A wealth of experience and expertise gained over the course of decades enables the NCC to conduct and supervise all kinds of mapping and spatial information projects at the national and international levels.

Private Aerospace Firms
Aerospace is a research-intensive industry, and due to the necessary capital outlays in research and development before the production of

output by this industry, only a few private technological firms operate in the country.

Infrastructure and Capacity Building

Capacity building is one of the key factors in the development of aerospace technology in any country. For Iran, it is also of great importance, and it is given the high priority it deserves.

Space Segment
In this section, the plans for technological capacity building and the development of expert human resources in the past few decades, which have resulted in many scientific and engineering achievements in the country, are discussed. The discussion focuses on the satellite programs that have been developed by Iran.

- **Zohreh**

Zohreh (which means "Venus" in Persian) is the satellite that was planned to meet Iran's telecommunications and Internet needs in the 1970s, and it was expected to be the first satellite developed in the country. Iran contracted with a Russian company, M. F. Reshetnev Scientific-Production Association of Applied Mechanics (the Russian acronym, NPO PM), in the Eastern Siberia town of Zheleznogorsk, to build the $132 million Zohreh satellite. It was designed to provide Iranians with numerous services, including television and radio broadcasts, the Internet, and email access. NPO PM technicians were supposed to assist their Iranian counterparts in controlling the system during its utilization, thus increasing its functionality (Nemets et al. 2009). The satellite was provisioned to have a lifetime of 15 years, using a Ku-Band frequency for receiving and transmitting, with an Alcatel and Astrium payload that included 12 transponders consisting of eight units of 36 MHz each and four units of 72 MHz each. In 2009, the Russian government, under international pressure, withdrew from supplying expertise and support for the advancement of the Zohreh project. Consequently, in September 2010, Iran announced that the satellite would be built and launched using domestic potentials and possibilities. The satellite is rescheduled to be launched in 2014 (Wikipedia 2012a).

- **Mesbah (Lantern)**

Building a satellite was what Iran had initially planned. Iran's Telecom Research Center (ITRC, or Sa Iran), a Defense Ministry-affiliated company, and the IROST affiliated with the Ministry of Science, Research and Technology, began developing the 130-pound [around 59-kg] Mesbah microsatellite in 1997 with the help of an Italian company, Carlo Gavazzi Space (CGSC). The Iranian Institute of Applied Research was the primary constructor of Mesbah. On August 4, 2005, Mesbah was showcased for the first time in an official ceremony. It was scheduled to enter orbit in early 2006 on board a Russian rocket. Once the Mesbah satellite was fully operational, personnel from ITRC/IROST planned to control it from a ground station in Tehran throughout its three-year lifespan. Mesbah would greatly expand Iran's understanding of space technology and create a solid base for further achievements in this area. However, by 2004 it appeared that the Mesbah project was troubled with multiple technical difficulties (Nemets et al. 2009).

Planning research began as early as 1997, and the prototype of the satellite was built during 1999–2001. Mesbah was the first satellite that was designed and manufactured in Iran. Having dimensions of 70×50×50 cm^3, the $10 million Mesbah satellite weighs 65 kg. It will fly 900 km over land, and will be controlled from a ground station located in the ITRC, while the back-up station will be operated by the CGSC in Milan. It will orbit the Earth 14 times daily, while being observable from Earth-based stations four times in 24 hours. Although the satellite has an expected life span of three years, it could be capable of continuing operations for up to five years. It is designed to cover Iran, but will be able technically to render services in Europe and the Americas (Zucconi 2005).

The responsibility for conducting and advancing the Mesbah project was given to the ISA as soon as the agency was established early in 2004 under the MCIT. Furthermore, the Space Technology Group of the Electrical and Computer Science Engineering Department at IROST actively conducts R&D in the satellite payloads and ground stations, as well as in space and aerospace applications. The Space Technology Group has used the systems-engineering design plan of Mesbah for its next satellite, Mesbah-2, for the ISA (IROST 2009).

- **Small Multi-Mission Satellite (SMMS)**

On April 22, 1998, under the Asia-Pacific Multilateral Cooperation in Space Technology and Applications (AP-MCSTA) initiative, Iran, China, South Korea, Mongolia, Pakistan, and Thailand signed a memorandum of understanding to build a SMMS. The main mission of the SMMS consisted of disaster and environmental monitoring, civilian remote sensing, and communications experiments (AP-MCSTA 2009).

The SMMS is a medium-Earth orbit Sun-synchronous satellite weighing 490 kg, with an orbit of 650 km above Earth. The total cost of the system is $44 million, and Iran's share of the cost of manufacturing and launching the satellite is $6.5 million. Having access to advanced Earth observation satellites is of great advantage for Iran because such an observation system has great utility, particularly after occurrences of natural disasters, for the coordination of relief efforts.

The SMMS carries a low-resolution, charge-coupled device (CCD) camera and an experimental telecommunications system that was built with Iranian technical contributions. The initial date of the launch of the satellite was 2004; however, delays pushed back the intended launch year to sometime in 2007. The satellite was launched by China and some other countries in 2008 (Shin 2009) under the Asia Pacific Space Cooperation Organization (APSCO), and is reportedly concluding its life span, as announced by the authorities of APSCO.

• **Sina-1**
Because of the multiple technical difficulties related to the implementation and advancement of the Mesbah project, the Iranian Institute of Applied Research requested Russian assistance to develop a microsatellite called Sina-1, which would meet all of Iran's desired characteristics and functions. The Russian Federation consequently developed the satellite and planned to launch it in September 2005. However, manufacturing delays in Sina-1 postponed the launch until October 27, 2005. With the launch of Sina-1, Iran became the forty-third member of the world space club (Nemets et al. 2009).

Sina-1, a 160-kg microsatellite, was launched by a Russian Kosmos-3M rocket from the city of Plesetsk in the Russian Federation to an altitude of 700 km. The satellite had a research mission related to remote sensing, that is, to monitor natural disasters and observe agricultural trends, and establish communications. The microsatellite is a Sun-synchronous near-polar orbiter with an inclination of 98.18 degrees and a period of 98.64 minutes. Iran acquired the satellite at a price of USD 15 million. The satellite, with dimensions of 80×130×160 cm^3, images Earth's surface from the Arctic to the Antarctic with a 50-m resolution in panchromatic mode and a 50-km swath. In multispectral scanning mode, the resolution is 250 m with a 500-km swath. The satellite transmits data on VHF and UHF frequencies, and is used in observing natural disasters, mineral resources, and crop yields. Iran had little input in the design and construction of Sina-1; however, the project has provided the country with valuable experience in ground-control tracking and telemetry handling (Schechter 2009).

• **Omid**

Iran's next major development in the space technology industry came to realization soon after early February 2008, when the country's first space research center was opened at the headquarters of Iran's Space Agency in Tehran, though the center itself is physically located in the northern Semnan province. This research center was used to launch Iran's first home-produced satellite "Omid" (Hope). Iran fired a satellite launch vehicle (SLV) that was designed to send the Omid satellite into orbit. Iran's state television channel announced that Iran is among the world's top 11 countries that have the technology to build satellites and launch rockets. The rocket blasted off from a launch pad in the Semnan province's desert terrain (Nemets et al. 2009).

On February 2, 2009, Omid, Iran's first domestically developed telecommunications satellite, was put into low-Earth orbit (LEO) with the country's first domestically manufactured SLV, named Safir-2. The launching of the domestically produced satellite and the SLV marked a great leap forward in space technology development for the country. The design, manufacturing, testing, and operation of the satellite, and the development of its launch vehicle were carried out by Iranian scientists and engineers (Iranian Space Agency 2009).

Completion of this project made Iran the ninth member of the exclusive club of countries that own independent satellite launching and manufacturing capacities, including the Russian Federation, the United States, France, Japan, China, United Kingdom, India, and Israel. The process of building Omid began in February 2006. It was a store-and-forward telecommunications 40-cm^3 satellite weighing 27 kg. With passive thermal control, it worked in the UHF band. The satellite's nodal period was 90.7 min with an inclination of 55.71°. Its apogee and perigee were 381.2 km and 245.5 km, respectively. Omid was supported by a network consisting of three telemetry, tracking, and command (TT&C) stations, one central flight control station, four ranging stations, and ground receiving stations and terminals. The mission of Omid ended on March 24, 2009, 50 days after its launch. The satellite entered the atmosphere in mid-April 2009, and it is estimated that the second stage of the Safir-2 had deteriorated before the end of May 2009 (Iranian Space Agency 2009).

• **Rasad**

In line with the R&D plans in space science and technology, the Iranian satellite Rasad, with a remote sensing mission, was launched on June 25, 2011, into an elliptic orbit of 350 km height at apogee and 240 km height at perigee. This satellite's main mission was to access the performance of its different subsystems and to provide images with appropriate resolution for meteorology and natural disaster management applications. This 75 kg satellite with a life span of 40 days accomplished its mission successfully (Shafti 2012).

Rasad, with 15.3 kg weight, orbited the Earth 15 times per 24 hours. Rasad was equipped with rechargeable batteries, with no limitation in the power source, and used its self-generating power capability. All stages of design, construction, project management, and testing of the Rasad satellite were carried out based on European Air Charter Safety Foundation (ACSF) standards. According to these standards, the subsystems are tested separately. Rasad's launch vehicle was a two-stage booster rocket, which placed Rasad in the elliptical orbit of 260 km trough drawdown. The rotation speed of Rasad at 286 km altitude from Earth was 48.7 km per second (Fazeli 2011).

• **Navid**

The ISA from its early days aimed at capacity building and encouraged active links between the agency and universities. This was accomplished by developing student satellites at a number of leading Iranian universities. Navid, a microsatellite of 50 kg weight was the first satellite built by the Science and Industry University of Iran (Elm-o-Sanat University of Iran) in conjunction with the ISA, and was launched on February 3, 2012. The satellite, with Earth observation as its mission, orbited around the globe in an elliptic orbit of 375 km height at apogee and 250 km at perigee six times a day, with a life span of two months (Shafti 2012).

Satellite Launch Vehicle Development

Developing SLVs is the result of close cooperation among academic institutions, the Ministry of Science, Research, and Technology, and the defense sector of the country, that is, the Ministry of Defense and Armed Forces Logistics (MODAFL). Moreover, bilateral cooperation with countries such as North Korea, China, and the Russian Federation has contributed to the success of the project. The availability of and access to domestically developed launch vehicles and rockets has played a key role in Iran's recent leap in space endeavors. To reach this important and strategic goal of acquiring rocket technology, Iran has made intensive efforts to prepare and make available launch vehicles, in keeping with its strong desire for self-sufficiency in this arena. The first successful example is the Safir-2 Launch Vehicle that Iran developed for orbiting its domestically developed satellite Omid. With a perigee and apogee of 250 km and 500 km, respectively, Safir-2 is 22 m long with a diameter of 1.25 m, and weighs more than 26 tons. The success of Safir-2 was due to experiments and achievements resulting from the successful launch of the suborbital rocket Safir on February 2, 2008, from Iran's domestic launch site in the northeast of the country. The launch of Safir, using Iran's Launch Vehicle (IRILV) was a preparatory mission for orbiting Omid (Tarikhi 2008a).

A joint threat assessment team consisting of experts from the United States and the Russian Federation examined Iran's nuclear and missile technological capabilities and offered a favorable assessment of Safir SLV. We find this judgment to be significant and wish to use it as an independent assessment of the degree of the technological sophistication of the Safir and Omid. "By launching an earth satellite, Iran has demonstrated that it can exploit low thrust rocket motors to build a two-stage rocket, and that it has qualified engineers who are able to make good use of the technology that is available to them. It does not show, however, that Iran has made a fundamental technological breakthrough." Furthermore, these experts declare that "the launch of the Omid satellite provides new information about the way in which Iranian rocket technology is developing. Iranian engineers have demonstrated a high level of competence and ingenuity in rocket design" (East-West Institute 2009).

Ground Segment
In addition to space technologies, Iran has been developing its ground facilities for communications and data acquisition throughout the country for many years.

After the Omid satellite was designed, work began on the design and construction of ground stations. The first communications command station was constructed in Charmshar near Tehran. Another one was set up in Chabahar, a coastal city on the Persian Gulf. A station for central positioning and control of the satellite was built in Semnan; four positioning stations were constructed in the Sistan and Baluchistan provinces; and a station for communications was built in Abadan.

It should be noted that these stations include tracking, telemetry, command, and a central station. The tracking station has the mission of tracking the satellite during the launch, and tracking and communicating with the satellite after it is placed into orbit. Using the received data from the tracking stations, the telemetry and command stations track the satellite and send commands to the spacecraft. The central stations have the responsibility of guiding and controlling the booster rocket, the satellite, and telemetry. Major achievements of this project are listed below.

1. Acquiring knowledge of the design and construction of ground stations;
2. Gaining the capability of developing the required software for the ground stations;
3. Acquiring the capability of establishing communication among the satellite ground stations;
4. Developing the capability of establishing a secure communication network among the ground stations.

Among the other achievements of this project, the construction of the first mobile remote testing station is noteworthy, which is the first receiving station of images acquired by a satellite that has been produced by Iranian experts. This station was constructed to receive data from remote sensing satellites that orbit between 800 km to 1200 km from Earth on X-band. The time for settling and preparing the station for work is two hours, and the station has the capability to receive, quickly observe, store, and produce images (Fazeli 2011).

Aviation Industry
The technological capabilities being developed in the aviation industry in Iran are mostly limited to technology transfers. However, the main achievements in this field consist of manufacturing the following items (Ebrahimi 2011):

- 4-seat and 2-seat aircraft and helicopters, 5,300-kg agriculture aircraft, gliders and paragliders and kites; unmanned aerial vehicle (UAV)s and remote control aircraft and helicopters;
- Radars for sea and air applications, laser and optical systems for airborne applications, satellite communication, and avionics systems;
- Composite and metal parts (Ebrahimi 2011).

The Civilian Aviation Industry
As stated earlier, the first initiative to develop the aviation industry in Iran was primarily for commercial purposes. However, subsequently, the idea of creating an air force attracted the attention of top policy makers in the 1930s This development necessitated the establishment of

technical and civil support organizations and auxiliary bodies to effectively implement the civil aviation services alongside the noncivilian services and logistics. Rapid developments in the industry facilitated a partnership between the private sector and Iran Airways, founded in 1944 through Iranian private investment. The Persian Air Service (PAS) followed the Iran Airways group, entering this arena in 1952. In 1961, Iran Airways and Persian Air Service merged to become Iran Airline. With the state nationalization of the air transportation industry in 1962, the Iranian National Airline (called Homa in Persian) was established and continued its activity under the regulations and mandates of the International Air Transport Association (IATA). Over the course of the years, the number of active air agencies has grown to 13, with 120 airliners, while Homa and Iran Air Tour are both state air agencies, and Aseman is active as a semistate agency.

The effort and attention that Iran has directed to the aeronautics and aviation industry for half a century, since the industry's inception in the 1930s, led to the establishment of a robust infrastructure and the emergence of capable and specialized human resources in Iran, and raised the country to the highest rank in the aviation industry in the Middle East, a position that Iran benefited from in later decades even after the advent of the Revolution in Iran in 1978.

Human Resource and Public Awareness

In the past two decades, Iran has worked on basic capacity building in space science and technology by developing education and training in this domain at the undergraduate and postgraduate levels. Fourteen universities and R&D institutions in the field of aerospace at the undergraduate and master's degree levels are currently active in the country. One aerospace research institute also provides education at the doctoral level in Iran. The subjects taught at these institutions of higher education include remote sensing, satellite telecommunications, and global positioning systems (Tarikhi 1994).

Aerospace engineering has been considered as one of the strategic careers in Iran in recent decades. The field was established for the first time in 1987 at Amir-Kabir University of Technology (AUT, formerly Polytechnique of Iran). In addition to AUT, Khajeh Nasir Tusi Technical University, Sharif Technical University, Imam Hussein University, Shahid. Sattari University, Malik Ashtar Technical University, and the Science and Research Branch of the Azad University, all of which are based in Tehran, presently offer aerospace courses at the bachelor's level, while the Tarbiat Modarres University, Elm-o-Sanat University of Iran (Science and Technology University of Iran), Mashed Ferdowsi University, Shiraz University, Shahid Beheshti University and the Aerospace Research Institute (ARI) offer aerospace courses at the master's in addition to the bachelor's level. In some of the top universities of Iran, including AUT, aerospace courses at the doctoral level are also available.

Employment Status of Aerospace Engineering Graduates
Graduates of the aerospace engineering universities are employed by the national defense industry (24 percent), the nondefense government sector (18 percent), and the private sector (27 percent). Fourteen percent of those who work for the private sector are employed in jobs related to aerospace, and 13 percent of these graduates in aerospace fields are involved in nonaerospace-related activities. Ten percent of the aerospace graduates leave the country for graduate studies or for employment. The remaining 12 percent either attend graduate school or are unemployed. These figures show that about two-thirds of 79 percent of the graduates in the aerospace field are employed by the government. Of the remaining one-third of aerospace graduates, 73 percent work for private firms, 14 percent are employed by private aerospace firms, and the remaining 13 percent work for nonaerospace private enterprises (Tabatabaeian 2008).

Educational Attainment of Aerospace Workers
Most aerospace workers have pursued postgraduate studies and are highly able engineers and scientists. Eighty-two percent of these employees have either master's or doctoral degrees. This figure refers to those who are employed by the domestic Iranian aerospace industry. This high level of educational achievement is an indication of these students' interest in advanced studies and the instructional staff's motivation for R&D. Moreover, the figure indicates that ample educational opportunity is available in this field in Iran (Tabatabaeian 2008).

In discussing the strong capabilities and skills of engineers and scientists who were trained in Iran and who are employed in the Iranian aerospace industry, the responses of Uzi Rubin, an Israeli engineer who managed Israel's "Arrow" program for missile defense, to a set of questions posed by Iran Watch[2] are reviewed below. In response to the question "What do you think about Iran's space program, and the recent rocket tests?" Rubin replies:

> I was impressed by the space launcher, especially its design. The first stage is a souped-up Shahab 3. The second stage is liquid fuel, but it's

storable liquid fuel. This is one step beyond basic, non-storable liquid fuel like what's used in Scuds. And it's not only storable, it's also hypergolic. That means you don't have to light the propellant, you just pump it into the motor and it spontaneously ignites. Iran managed to design a very elegant second stage—and also very light. The overwhelming majority of countries starting out with space launch technology started with three stages, why? Because it's easier to reach orbital velocity with three stages than with two stages. Doing it with only two stages places very stringent requirements on the second stage. But Iran did it. Everyone was surprised. This is something we didn't expect. I'd say it was an audacious achievement for a starting country. (Iran Watch 2009)

Furthermore, in response to the question "Were you surprised that Iran had mastered staging? Staging is not easy," Rubin gives the following assessment:

Well, staging is challenging, but I'm not surprised. When Iran announced its space program back in 1998, this obviously meant they were going to rely on staging. So I wasn't surprised that they did it. Staging is key to a space program. I was surprised that it worked so well the first time.

I was also surprised that Iran progressed so quickly. In February 2008, Iran fired a missile it called Kavoshgar, which was a Shahab-3, probably with a slightly stronger motor and painted in blue and white because it was flown by the space agency, not by the military. It had a typical triconic front end.... but with some changes. That was the first launch. Iran claimed the test was successful, but we saw it as a dismal failure. Six months later, in August, Iran fired its first space launcher [Safir]. No pieces fell off. That indicates ample telemetry data from the failed test, which allowed Iranian engineers to figure out what went wrong in February. They then displayed good engineering: they fixed the problem. Finally, the short recovery time—only six months—indicates vigorous program management.... as an engineer, I take my hat off. (Iran Watch 2009)

Global and Regional Cooperation

The ISA has given the highest priority to international cooperation based on the policy provisioned as one of its missions. In light of this goal, the ISA's main contribution to this subject is its close cooperation with the United Nations agency COPUOS. It is important to note that the ISA's representatives contributed to the COPUOS Bureau's work in the capacity of the Second Vice-Chairman and Rapporteur from 2004 to 2006, and the Chair of the Legal Subcommittee of COPUOS from 2010 to 2011. Moreover, Iran has contributed to the implementation of the

Recommendations of the Third United Nations International Conference on the Exploration and Peaceful Uses of Outer Space (UNISPACE-III) by chairing its Action Team number 1, which focused on the Development of a Worldwide Comprehensive Strategy for Environmental Monitoring which has continued since 2001 (UNOOSA 2009). Additionally, the ISA has taken part in creating a UN Space-Based Platform for Disaster Management (UN-SPIDER) and establishing one of its regional support offices in Iran. Furthermore, the constructive deliberations of the Iranian delegations on various space-related issues from technical, scientific, and legal perspectives provide strong evidence of Iran's interest in actively working toward the use of space for peaceful purposes.

The ISA has organized and hosted a number of workshops and seminars related to space science and technology applications, with special emphasis on remote sensing and disaster monitoring and mitigation in cooperation with UN-OOSA, the Inter-Islamic Network on Space Technology (ISNET), and other global and regional organizations. At the regional level, Iran cooperates with the UN Economic and Social Commission of Asia and the Pacific (ESCAP) and follows the plans developed by its Regional Program on Space Technology Applications (RESAP). ESCAP and Iran worked closely for establishing a Center for Informed Space-based Disaster Management and an affiliated research center (ESCAP 2006). On the initiative of the Asia-Pacific Multilateral Cooperation in Space Technology and Applications (AP-MCSTA) Iran cooperated in manufacturing a small multimission satellite for disaster management. Moreover, Iran joined the Convention of the Asia-Pacific Space Cooperation Organization (APSCO) under AP-MCSTA on October 28, 2005, as one of its founders and signatories along with China, Indonesia, Pakistan, Thailand, Bangladesh, Mongolia, Peru, and Turkey (AP-MCSTA 2009).

It should be noted that there exists considerable, growing interest in space science and technology in the Asia Pacific region. The big players are countries like China, India, Japan, and Australia, while South Korea, Thailand, Pakistan, and Iran are advancing rapidly. Iran is also a member of another space-related organization, the Committee on Space Research (COSPAR) of the International Council of Scientific Unions (ICSU). Activity and membership in the International Society of Photogrammetry and Remote Sensing (ISPRS), the Asian Association on Remote Sensing (AARS), the Asia-Pacific Satellite Communications Council (APSCC), among others, are all indicators of Iran's desire to stay active in the international cooperation programs in space applications and science. Furthermore, it is noteworthy that nearly all of Iran's endeavors to implement its satellite projects such as Mesbah, Zohreh, SMMS, Sina-1, and Omid, as well as the establishment of the MSRS and the installation of other space ground stations and facilities, have involved international cooperation (Tarikhi 2009).

The Investments

Investment in the aerospace industry in Iran is primarily a public investment, and private investment in this field is rare. The reason for the scarcity of private investment in this industry is due to the lack of knowledge on the part of the general public about the size of required outlays in this industry. Generally, because of national security concerns, the amount and composition of investment in aerospace activities in Iran are not publicly known. This lack of information has led to few, if any, private-sector investments in this field. However, the Chairman of the Board and the Chief Executive Officer of the Iranian Aviation Industries Organization have stated that the total planed investment in this industry during the Fifth Development Plan (2011–2015) is $1 billion (Transportation Industry Network 2010).

Analysis of the Current Conditions of Iranian Aerospace Industry

In this section, taking a broad perspective and relying on existing studies, we analyze the present conditions of the aerospace industry in Iran and identify the capabilities, weaknesses, challenges, and impediments facing the industry.

The Weaknesses

Not surprisingly, a review of the status of the aerospace technology in Iran indicates that the domestic capabilities are lagging behind the technologies in advanced industrial countries. A number of factors could be contributing to this gap. These factors are not unique to the aerospace industry in Iran, but rather are the hallmarks of difficulties in industrial development in all technologically latecomer countries. We enumerate the following as the potential contributing factors: (Tabatabeian 2009)

- International sanctions and difficulties in procuring technologies, instruments, and parts from abroad;
- Inadequate capabilities in the design of products and the replication of productive processes;
- Inadequate process innovation;

- Lack of innovation in general
- Protection of intellectual property;
- Low level of support and marketing;
- Weakness in the commercialization of existing products and processes.

The Institutions
Most institutions in this sector, both from a standpoint of policy making and R&D of technology fields, are not in a suitable condition. A limited number of the research centers in this field usually face the problem of inadequate interactions and communications among each other, and they tend to perform duplicated tasks. Of course, this problem emerges from the lack of a centralized decision-making system.

Human Resources
The engineers and scientists who are employed in this industry have received independent, international acclaim from American, Russian, and Israeli experts, as noted previously. However, there is room for improvement, most notably in the need to close the chasm between the academic knowledge acquired by students and the practical knowledge required by the industry. Currently, there are five tracks in the fields of aerospace engineering, including structure, aerodynamics, flight mechanics, propulsion, and flight engineering. Much of the required expertise, including electronics and communications, is acquired from other academic disciplines, which leads to less than optimal recruitment (Tabatabaeian 2009).

Resources Allocation and Investment
Because of the public allocation of funds, which constitute an overwhelming part of the total capital outlays in this field, investment in this sector is very stable. However, due to the lack of a centralized decision-making body, investment funding is not prioritized, and gross inefficiency occurs in the form of investment in duplicated projects. Nevertheless, recent annexation of the space agency and its affiliated organizations and research institutes to the Presidency Institution of the IR of Iran is a remarkable improvement in concentrated policy making, the management of projects, and the allocation of funds in this field.

Future Trends and Plans

The main goal of the Iranian government is to advance space technology through the processes of sending a human being into space, and by acquiring knowledge in the design, manufacturing, and launching of a satellite into the Earth's orbit in collaboration with other Islamic countries. The first step in implementing this idea was the development of a General Plan for Development of Aerospace Science and Technology for the country. The main objectives of this General Plan are as follows (Supreme Council of Science, Research and Technology 2008):

- Purposeful, balanced development of those branches of aerospace science and technology, which are harmonious with national advantages and capabilities;
- Using the aerospace industry as one of the main sources of the acquisition of knowledge and wealth for increased national welfare;
- Improvement of national security by developing aerospace science and technology in collaboration with other countries in the region;
- Establishing an effective presence for the private sector in national and international aerospace activities;
- Increasing the global competitiveness of domestic aerospace research and technology centers;
- Increasing the productivity of experts in producing aerospace science and technology;
- Improving the capabilities to design and manufacture aerospace products in order to meet national needs and be present in the international arena.

Iran's Manned Space Flight Program

The background of Iran's manned space flight program dates back two decades. The country revealed its intention to send a human into space on June 21, 1990, in the course of the summit of the presidents of Iran and the Soviet Union. Both countries agreed to make joint Soviet-Iranian manned flights to the Mir space station. The dissolution of the Soviet Union soon after in 1991 led to the interruption of the agreement. However, in November 2005, the authorities of the ISA announced a plan for manned space flight and plans for the development of a spacecraft and a space laboratory as well. On August 20, 2008, Iran made known the country's plan to launch a manned mission into space within a decade, for which the goal of making Iran the leading space power in the region by 2021 was divulged (Tarikhi 2011c).

Since the successful launch of Iran's first domestically made telecommunications satellite, Omid, with Safir-2, the country's first domestic SLV on February 2, 2009, Iran has started a new program for designing, manufacturing, and putting new satellites into orbit. Although Omid was an experimental satellite with a short-term mission of taking orbital measurements, the experience and knowledge gained through orbiting and operating it opened the door to setting up more sophisticated systems that carry Earth observation apparatus, and communication and research tools. Iran has begun practical experiments on life in space by developing a space biocapsule in line with the country's plan to send astronauts into space in 15 years. In 2010, Iran developed a more powerful launch vehicle named Simorgh (Phoenix, in Persian) with the mission of carrying heavier satellites into orbit.

However, there is another way to get into space, which allowed an Iranian–American telecommunications tycoon businesswoman to travel into space. Anousheh Ansari, who spent her childhood in Iran dreaming of venturing into space, realized her dream when she boarded a Russian Soyuz spacecraft to fly into space. She became the fourth person—and the first woman—to pay an estimated $20 million to travel into space. Wearing both the American and Iranian flags on her spacesuit during training, Ansari stated that she wanted to recognize both countries' contributions to her life (Harvey et al. 2010). She went on a ten-day journey to the International Space Station as part of a crew-exchange flight starting on September 18, 2006, from Baikonur, Kazakhstan, and landed safely aboard Soyuz TMA-8 on September 29, 2006 (Ansari et al. 2010).

Conclusions and Recommendations

Iran's involvement in aviation activities dates back nearly a century. In light of the country's top authorities continuing to devote attention to this field, a strong and robust infrastructure in the aeronautics and aviation industry was established that gave Iran the highest ranking in this field in the Middle East. However, Iran's involvement in space activities originally began in the form of its participation in creating the United States' satellite telecommunications network, by its allowing the construction of a satellite data-receiving station in Iran in the early 1970s. Iran's serious national efforts to use and develop space technologies for peaceful purposes did not begin until the early 1990s, after a decade of revolutionary upheavals and the war with Iraq. The design and development of satellites like Zohreh, SMMS, and Mesbah are examples of the country's postwar national R&D efforts in acquiring space technologies that took place mainly under the auspices of the Ministry of Science, Research, and Technology, the Ministry of Communications and Information Technology, and the Ministry of Defense and Armed Forces Logistics.

The manufacturing of Mesbah was the result of cooperative efforts between the Ministries of SRT and CIT, MODAFL, and the Italian space firm CGSC. The design and manufacturing of the SMMS took place within a multilateral framework at the regional level and involved Iran, which had the aim of disaster monitoring. In this process of development, the Zohreh satellite project represents another outcome of capacity-building efforts in space technologies, which was a concern of Iranian authorities for decades. It is clear that Iran's civilian space industry made major achievements and major contributions to providing the country with independent access to outer space and developing satellite and communication systems, as well as systems for remote sensing and navigation.

Of course, analysis of the current situation of aerospace R&D reveals some obstacles to Iran reaching its goals. To remove these obstacles, improving the efficiency of policy-making bodies and institutions; developing the necessary infrastructures, including expansion of the educational and training opportunities in aerospace-related sciences and technologies; offering multidisciplinary courses and degree programs relating to aerospace; strengthening networking and promoting international cooperation; employing competent and eligible aerospace managers and technocrats; and promoting private-sector involvement and investment in the industry are recommended.

To attain its scientific goals for improvement of the national well-being, the country relies on its own human resources and technological capabilities. Iranian technical capabilities in space sciences and exploration are increasing rapidly. Investing in space is expensive. Such high investment can be justified only if all the achievements of the country's space program are integrated into the social, economic, educational, and technical life of the nation. Iran's entry into space exploration and its peaceful applications, by relying on the country's indigenously developed technological system, has elicited a notable, unprecedented pride among its citizens. Iran's advances could potentially provide a considerable impetus for further productive cooperation in space between Iran and other nations. International cooperation, similar to what already exists within the framework of COPUOS activities on the peaceful uses of outer space, could improve the world community's understanding of Iran's space policy and its visions for the twenty-first century (Tarikhi 2009). Following the approval of the Iranian Administrational Supreme Council in late September 2010, the ISA, the pivotal player in Iran's space endeavors, was annexed to Iran's Presidency Institution, and consequently, the activities of the ISA under the Ministry of Communications and Information Technology were concluded after approximately seven years. This new organizational upgrading is an indication of how significant Iran's space endeavors are for the country's top authorities. However, it is imperative to take the necessary actions in conformity with legal and administrative requirements and obligations, and most importantly, to set up a newer statute based on new administrative changes to be approved by the Iranian Parliament. The new statute needs to clarify the relationship between the space agency and the Supreme Space Council (SSC), which was revived with the approval of the Expediency Council in September 2008. Through this annexation, in the meantime, the ISA has begun efforts to create a new configuration for the agency that would be crystallized by the ISA's new organizational chart being approved. Reportedly, this reorganization places the greatest significance on technology development and research in the engineering of space systems, including satellites, manned space flights, space probes, space-related sciences, and ground-based launching platforms.

In addition to the development of a stronger launch vehicle, the country has been witnessing a boom in the design and development of new satellites. The research microsatellite Navid, developed by the Tehran-based Elm-o-Sanat University with the mission of observing Earth (remote sensing), is the latest to be orbited; new satellites like AUTsat, Tolou, Mesbah-2, Fajr and others are on the way. The quantitative growth of the Iranian satellites developed after the launch of Omid, in comparison to those that were planned before it, is estimated to be tripled. In recent years, much of the state funds have been allocated toward persuading the academic and noncivilian sectors to put resources jointly or individually into designing satellites and launchers/rockets. The rise in fund allocation has led to a quantitative increase in plans and projects for designing and manufacturing satellites. These efforts in development essentially require concentrated policy making, and economic and efficiency caretaking of their missions and functions, which consequently indicate the need to place supervision and control of the national space agency under the auspices of the SSC. All legislative, administrative, managerial, and executive efforts play a vital role in this respect, while each is affected individually or synergistically by the attitudes and visions of the authorities and policy makers, which could be discouraging or encouraging (Tarikhi 2011a). The more these visions and attitudes tend toward realism and open-mindedness, the more advances in space technology development will be successful and triumphant.

Notes

1. For a definition of latecomer firms, please see introduction of this book.
2. Wisconsin Project on Nuclear Arms Control is a Washington, DC-based organization that "carries out research and public education designed to stop the spread of nuclear weapons, chemical/biological weapons and long-range missiles." Iran Watch is a publication of this Wisconsin project. For more details about the organization, visit http://www.iranwatch.org/aboutus/.

References

Aerospace Research Institute. n.d. "About Institute." Ari.ac.ir. Accessed August 17, 2011. http://www.ari.ac.ir/.
Aerospace Technology Development Administration. 2009. "Aerospace Technology Development Administration Introduction." Aero-space.ir. Accessed August 2011. http://www.aero-space.ir/.
Ansari, Anousheh, and Homer Hickam. 2010. *My Dream of Stars: From Daughter of Iran to Space Pioneer.* New York, Palgrave McMillan.
Asia-Pacific Multicultural Cooperation in Space Technology and Applications (AP-MCSTA). "SMMS for Environmental Monitoring and Disaster Management." Apmcsta.org. Accessed May 11, 2009. http://www.apmcsta.org/Projects/SMMS-Management.aspx.
———. "APSCO Motives." Apmcsta.org. Accessed May 11, 2009. http://www.apmcsta.org/Apsco/motives.aspx.
East West Institute. 2009. I"ran's Nuclear and Missile Potential: A Joint Assessment by U.S. and Russian Technical Experts." Ewi.info. Acccessed May 11, 2009. http://www.cwi.info.
Ebrahimi, I. 2011. "Iranian Aviation and Space Industries Association." Personal interview with M. Abbassi and M. Ashrafi. August.
Electrical and Computer Science Engineering Department. Iranian Research Organization for Science and Technology (IROST). Accessed May 10, 2009. http://www.irost.org/en/computer/index.asp?code=2.
Fazeli, Hamid. 2011, "Iran Satellites." Personal interview with M. Abbassi. September.
Harvey, Brian, Henk Smid, and Theo Pirard. 2010. *Emerging Space Powers; the New Space Programs of Asia, the Middle East, and South America.* Chichester, UK: Springer/Praxis.

Implementation of the Recommendations of UNISPACE III. 2003. UNOOSA. Vienna, Austria. Accessed January 20, 2012. http://www.unoosa.org/pdf.

Iran Supreme Council of Cultural Revolution. 2010. *Iran Science and Technology Roadmap*. Tehran: Iran Supreme Council of Cultural Revolution.

Iran Watch. 2009. "Iran's Missile Program Making Steady Progress: An Interview with Uzi Rubin." Iranwatch.org. Interview Series. Accessed January 30, 2012. http://www.iranwatch.org/ourpubs/roundtables/interview-rubin-091709.htm.

———. "Iran's Missile Program Making Steady Progress: An Interview with Uzi Rubin." Iranwatch.org. Accessed January 12, 2012. http://www.iranwatch.org/ourpubs/roundtables/interview-rubin-091709.htm.

Iranian Aerospace Society (IAS). n.d. Ias.ir. Accessed December 15, 2011. http://www.ias.ir.

Iranian Research Organization for Science and Technology (IROST). n.d. Irost.org. Accessed Februart 12, 2012. http://www.irost.org.

Iranian Space Agency. n.d. "Iranian Space Agency Statute." Isa.ir. Accessed August 20, 2012. http://www.isa.ir/index.php.

Islamic Republic of Iran Inventory. 2006. "ESCAP2006." Unescap.org. Accessed January 30, 2012. http://www.unescap.org.

Nemets, Alexandr V., and Robert W. Kurz. 2009. "The Iranian Space Program and Russian Assistance." *Journal of Slavic Military Studies* 22: 87–96.

Schechter, Erik. 2009. "Aerospace Industry in the Middle East." *Aerospace America Magazine*, April, 39–40.

Shafti, Hassan. 2012. "Statement by the Head of Delegation of the Islamic Republic of Iran, to the 49th Scientific and Technical Subcommittee Meeting of the Committee of the Peaceful Uses of Outer Space (COPUOS)." February 6–17, 2012, Vienna, Austria. Accessed 20 October 2012. http://water-portal.com/index.php?option=com_content&view=frontpage&Itemid=1.

Shin, Jenny. 2009. "A Chronology of Iran's Space Activities." Center for Defense Information (CDI). Accessed January 6, 2012. http://www.cdi.org/pdfs/IranSpaceTimeline09.pdf.

Supreme Council of Science, Research and Technology. 2008. "Aerospace Science and Technology Development Document." Supreme Council of Science, Research and Technology, Government of the I.R. of Iran.

Tabatabaeian, S. H. 2008. "An Analytic Report of Iran Aerospace Sector." Tehran: Research Deputy of Ministry of Science, Research and Technology.

———. 2009. "Iran Situation in Aerospace Science and Technology." Tehran: Technology Deputy of Ministry of Science, Research and Technology.

Tarikhi, Parviz. 1994. "Sanjashazdordar Iran darchemarhaleiaz rushed hast?" (What is the Growth Status of Remote Sensing in Iran?) *Iranian Remote Sensing Center Newsletter*, 4, (1): 8, 12–13.

———. 2003. "Milestones of the Establishment of the Iranian Space Agency." Tehran: Iranian Remote Sensing Center (IRSC).

———. 2006. "Roads to Prosperity (part 1)." *GEO-connexion International Magazine*, 6 (December 2006/January 2007): 34–35.

———. 2007. "Roads to Prosperity (part 2)." *GEO-connexion International Magazine*, 6 (February 2007): 52–53.

———. 2008a. "Iran's Space Ambitions Ride on Safir." *Asian Surveying and Mapping (ASM) Magazine*. Accessed December 31, 2011. http://www.asmmag.com/features/729.

———. 2008b. "Mahdasht Satellite Receiving Station Verging into a Space Center," A blog on the legal aspects of human activities using aerospace technologies. Accessed December 31, 2011. http://rescommunis.olemiss.edu.

———. 2008c. "On the New Iranian Space Law Res Communis," A blog on the legal aspects of human activities using aerospace technologies. The University of Mississippi School of Law. Accessed December 31, 2011. http://rescommunis.wordpress.com/2008/07/09/guest-blogger-parviz-tarikhi-on-the-new-iranian-space-law.

———. 2008d. "Statutes of the Iranian Space Agency." *Journal of Space Law* 34 (2): 487–495.

———. 2009. "Iran's Space Programme: Riding High for Peace and Pride." *Space Policy International Journal* 25 (3): 160–173.

———. 2010. "More Significant Role for Iran's Space Administration," A blog on the legal aspects of human activities using aerospace technologies. Accessed February 27, 2012. http://rescommunis.wordpress.com/2010/11/11/more-significant-role-for-iran%E2%80%99s-space-administration/.

———. 2011a. "Iran's Space Development: To Reach the Zenith Let's Fly Altogether as a Simorgh," A blog on the legal aspects of human activities using aerospace technologies. Accessed March 18, 2012. http://rescommunis.wordpress.com/2011/02/07/guest-blogger-parviz-tarikhi-iran-space-development-to-reach-the-zenith-lets-fly-altogether-as-a-simorgh/.

———. 2011b. "Supreme Council of Space in the Way to Play its Supervisory Role," A blog on the legal aspects of human activities using aerospace technologies. Accessed March 18, 2012. http://rescommunis.olemiss.edu.

———. 2011c. "Manned Space Flight Mission of Iran." Accessed March 10, 2012. http://rescommunis.olemiss.edu.

Telecommunication Company of Iran. 2008. "Technical Report." February 9.

Transportation Industry Network. 2010. "Aviation Sector." Transportation Industry Network News. Accessed January 30, 2012. http://tinn.ir/Air_Sec.

United Nations Office for Outer Space Affairs (UNOOSA). 2009. Accessed May 7, 2011. http://www.unoosa.org/oosa/en/COPUOS/copuos.html.

Wikipedia. 2012a. "Iranian Space Agency." Accessed January 31, 2012. http://en.wikipedia.org/wiki/Iranian_Space_Agency.

Wikipedia. 2012b. "Iran Aviation Industries Organization." Accessed March 12, 2012. http://en.wikipedia.org/wiki/Iran_Aviation_Industries_Organization.

Zucconi, L. 2005. "The Mahdasht Project." Carlo Gavazzi Space SpA. Accessed December 30, 2010. http://www.itrc.ac.ir/ist2005/Keynote/K5/MESBAH-Conf.pdf.

CHAPTER 10

The Automotive Industry: New Trends, Approaches, and Challenges

Manochehr Manteghi

Introduction

Industrial development is one of the major factors in economic growth. In today's competitive world, the manufacturing industry has a strategic position in development processes because industrial development, in addition to increasing output and employment, facilitates the provision of services and increases productivity in the other sectors of the economy.

Among industries, the automotive industry as a "key" economic sector with 120 years of history plays an increasingly fundamental role in human societies. Also, with more than seven hundred million vehicles in the world, the automotive industry has changed the economic and cultural dimensions of societies profoundly and extensively. The automotive industry is comprised of an array of goods and services such as metals, plastics, chemicals, insulation, glass, electrical components, electronics, metallurgy, and design that involve the arts and reflect cultures, as well as the management of technologies. Therefore, innovations in all of these products and services have been influential in the design and production of automobiles and have contributed to the development of this industry as one of the largest value chains and sources of employment.

As such, it must be acknowledged that the automotive industry plays an important role in terms of employment creation, the use of advanced technologies, and the generation of profit. Moreover, the cultural and societal impact of the industry, such as consumer welfare, freedom of choice, entertainment, and sports, are notable. At the same time, one should keep in mind the cyclical nature of the industry as well as the strong dependency of the industry on governmental support.

Below we will trace the development of the Iranian automotive industry from its very inception to 2011.

Background of Automotive Industry in Iran (1958–1996)

In this section, we briefly review the path of development of Iran's automotive industry from the beginning (1958) to 1996. This is the period when the first steps were taken toward manufacturing automobiles in the country. This period can be referred to as the "assembly" period. The later period between 1996 and 2011, which is considered the coming of age of automobile manufacturing in Iran, will be discussed in below in the section titled "Actors in the Iranian Automobile Industry."

The development of the automobile industry in Iran, from its inception in 1958 until 1996, is punctuated with frequent ups and downs that were mostly caused by radical political changes and instability in the country. The main activities of the industry during this period, which constitutes more that 70 percent of the life of the industry, was primarily dedicated to the assembly of imported manufactured parts. The country's economic dependence in general and the auto industry in particular on the West in the pre-Revolution era; its heavy reliance on oil revenues instead of manufacturing; the large imports of automobiles and auto parts; the eight-year war with Iraq (1980–1988); and automakers' lack of initiative to innovate, which emanated from a noncompetitive business environment created by protective tariffs, collectively caused frequent and successive delays in the growth and development of the automotive industry. In the meantime, in spite of the presence of these unsettling conditions, the government had no industrial policy for the automobile sector of the economy.

In this period, the import of automotive parts was dependent on the country's foreign exchange earnings from the oil exports. When the foreign exchange earnings of the country rose because of higher oil revenues, the import of auto parts increased, and when the oil revenues declined, so did the import of auto parts. The dependency of the automobile industry on petroleum exports is tantamount to the unfortunate loss of many opportunities for the industry to develop

In contrast, for example, South Korea started its automotive industry at roughly the same time as the automobile industry began in Iran, but it went through the stages of development much faster, entered the globalization stage, and began to supply international markets with high-quality and competitive products (Manteghi 2011). What follows is an overview of different stages of the development of the assembly period.

No Mileage (1958-1978)

With the founding of Jeep factories (now Pars Khodro) in 1958, Iran's automotive industry took the first step in the assembly of auto parts. Within a decade, a number of car-assembly plants under licenses from foreign automakers were set up in the country. In 1960, Saika Industrial Company was established, and it began to assemble the Fiat 1100 under the license of Fiat of Italy. In 1962, another automobile-assembly firm, named Iran Khodro, also known as Iran National, which is a public joint-stock company, was established. Iran Khodro was licensed by the British Rootes Group to assemble Hillman Hunter automobiles with the Iranian brand name of Paykan in Iran.

Iran Khodro produced only about 30 percent of the auto parts domestically. Furthermore, in 1968 another automobile-assembly company, Saipa, licensed by French automaker Citroen Company, began assembling Citroën's two-cylinder mini passenger car, the Dyane, in Iran. This car was named the Zhian.

Pickup truck production in Iran started in 1969 with Iran National's manufacturing of Paykan pickup trucks. Saipa Company introduced pickup trucks to the market in 1975. In 1969, Zamyad Company started assembly of the Nissan pickup under the license of Nissan Company of Japan. Also, Iran Kaveh Company was a manufacturer of General Motors Company (GMC) pickups between 1969 and 1977. Shahab Khodro Company (formerly Leyland Motor) manufactured 4- and 2- wheel drive pickups in limited numbers (up to 150 units) between 1969 and 1977. Iran Vanet Company (formerly Mazda, and now Bahman Group) started manufacturing Mazda 1000 and Mazda 1600 pickup trucks in 1970 (Parsa 2009).

Production of commercial vehicles in Iran started with the establishment of Khavar Company (under a license from Mercedes Benz Company) in 1959. Companies that produced trucks included the Zamyad Company (under a license from Volvo), Iran Kaveh (under a license from Mack), and Leyland Motors (under a license from Leylands). All of these truck-manufacturing companies were founded between 1963 and 1965.

In 1963, the Ministry of Economy issued new regulations for the issuance of permits for automotive production. For the first time, the Ministry adopted an industrial policy regarding the automobile industry that rejected assembly production based on total reliance on imported parts and limited the issuing of permits to applicants who were able to at least manufacture the body, the floor, and all the parts that involved compressing metals. Moreover, these applicants were required to utilize all domestically produced auto parts.

The idea of domestic production of parts was pursued for the rest of the decade of the '60s in Iran, and some investments were made to achieve this goal. However, these investments led to the production of only a limited number of items that were intended to be produced, and the work could not be completed until the first few years of the next decade. Among the companies that were established through these investments and that produced auto parts domestically, were the Iran Radiatorsazi Company (established in 1963 and started operations in 1976), which manufactured different kinds of radiators, and the Charkheshgar Company, (established in 1963 and started operations in 1975), which manufactured part of the drive system. Also in 1967, the Paykan engine assembly facility started its operations, with an annual capacity of sixty thousand units. This company directly and indirectly facilitated the production of many basic automotive parts inside the country. For example, in 1974, Iran Khodro's casting and motor factory was established to produce 6 parts of Paykan's engine. In addition to the investments in Iran Khodro, other investments were made too. As another important example of the auto part-producing activities of Iran Khodro, we cite the setting up of a factory for the production of wheel rims, exhaust systems, and parts of the brake system.

The auto industry in Iran was established and began operations at a time when two medium-term import-substitution strategies to industrialize the national economy were being partially implemented. These development plans consisted of the Third National Development Plan (1962–1967), with the strategy of the domestic production of imported consumer goods, and the Fourth National Development Plan (1968–1973), with the aim of the domestic production of imported intermediate and capital goods, the goals of which were partially met. Incomplete realization of the goals of the Third and Fourth Development Plans amounted to the country taking rudimentary and limited steps to find the lost links in the long chain of the industry, from assembling to part manufacturing, in the process of realizing self-sufficiency in automobile technologies.

It should be noted that industrialization as the cornerstone of development of the economy during Shah Reza Pahlavi's regime was set during the Fourth and Fifth Development Plans (1973–1978). However, for reasons examined below, these policies failed to establish an independent, healthy economy. Nevertheless, the automobile industry, along with other industries such as steel, petrochemicals, aluminum smelting, machinery, tractor and heavy machinery, and defense were established during these two planning periods. It should further be noted that in contrast to the other industries that were established by government investments, investments in the automobile industry were mostly private capital outlays.

The time between 1969 and 1977 may be considered as the period in which the automobile industry entered into a new, more advanced phase in Iran. During this time, the industry received weak and incomplete governmental support. In the first 5 years of this period, nominal support without a harmonious plan resulted in the encouragement of private investments in industrial enterprises, the enactment of laws to limit imports, and the protection of private investments. However, beginning in 1973, because of the rapid rise in the price of petroleum, which led to high inflation and the increased import of motor vehicles from abroad, the growth and development of this nascent industry faced stumbling blocks.

Moreover, in addition to these macroeconomic effects, two reasons at the firm level are cited for the inadequate progress in the auto sector of the economy. First, an insufficient number of enterprises for the production of parts, as well as the semitraditional, semi-industrial world outlooks of owners and managers of the automobile firms, are said to have had a retarding effect on the development of the industry (Manteghi 2011).

Furthermore, during this time the Pahlavi regime's support of industrial investments consisted of allocating bank credit and granting of permits to industrialists to open factories. However, because private investors, who often directly managed the enterprises they owned lacked expertise in formulating strategies to develop automobile technologies, no particular strategy then existed for expanding the automobile industry. Other factors that contributed to the lack of a strategy for developing the automotive industry in Iran included a lack of experts in advanced auto technologies and the government's failure to come up with a strategy to use the automobile industry as the core industry for industrial development of the economy as whole. Additionally, the most important reason for lack of development of this industry was the total reliance of domestic firms on transnational enterprises, in which the transnational companies were not readily inclined to assist in the independent development of the technological capabilities of the Iranian firms.

The primary policy of the Iranian government during the Fourth Development Plan (1968–1973) was to transform Iran's agricultural-oriented economy to a manufacturing-oriented one. During this period, the government established large industrial units, which were listed above, while simultaneously, it left the creation of assembly plants for the production of automobiles, among other consumer goods, to private-sector investors. Moreover, because of the auto industry's technological dependency on transnational companies, the automobile firms had to completely rely on the import of auto parts such as steel sheets, engines, and the other sensitive parts for production. The government's policy in fact encouraged such a technological dependence through the allocation of subsidized foreign exchange for an ever-increasing import of auto parts.

It is important to draw the reader's attention to the primary motive of Iranian investors in the automobile industry and their international partners. These entities were mainly motivated to realize the maximum profit in the shortest possible time. In the environment of crony capitalism that existed at the time, and with the motive of maximizing profit in the shortest time, one can clearly deduce that the industrialists at the time did not have any desire to acquire technological capabilities, which entailed the know-how that required risky investment in R&D and a long-term gestation period.

It should be mentioned further that one of the aims of the Sixth and last development plan (1976–1981) under the Pahlavi rule in Iran was the further development of the automobile industry, a plan that ended at the onset of the Revolution in 1979.

Struggle for Survival (1980–1989)

During the revolutionary upsurge and the general strikes in 1978, auto production decreased, and economic instability as well as capital flight increased. As a result, in 1979, to prevent further capital flight, and based on Article 44 of the new Iranian Constitution, all basic industries, including the automobile industry, were nationalized. Furthermore, according to the Law of Protection and Development of National Industries, the automobile companies were placed under the control of the National Industrial Organization (NIO), a government organization that was set up to manage large enterprises that were either bankrupt or were on the verge of bankruptcy because of the revolutionary upheaval in the country. The trend of the NIO controlling the main industries was completed by the time of the imposition of the US economic sanctions on Iran in early 1980.

In the midst of the revolutionary changes, the import of vehicles was prohibited in 1978, at which time the import of cars dropped from its peak number of 65,548 in 1977 to 32,193 units in 1978, to a small number of 3,291 vehicles in 1979, and a meager 667 units in 1980.

Shortly after the triumph of the Iranian Revolution, the strategy of producing automobile and auto parts in Iran assumed a peculiar place. With the start of the Iran-Iraq war in September 1980, a difficult economic period.

The car-assembly companies in Iran, by relying on the inventories of the parts in warehouses as well as procurement from suppliers other than their transnational partners, continued assembling cars during 1979–1981. Moreover, the rapid increase in auto production during 1974–1977, which took place mostly by importing automobile parts and without expanding the domestic capacity to produce auto parts, was not sustainable when the country faced an acute shortage of foreign reserve, trade embargos, and a restricted capacity to produce parts domestically. As a result, domestic car production during this time fell drastically. In general, with the exception of 1983 and 1984, when Iran's foreign exchange earnings rose because of the higher price of petroleum, during the years of war with Iraq, the automakers, which faced restricted access to subsidized foreign exchange, could not acquire parts from domestic suppliers (the domestic auto-part companies could not supply adequately) and were forced to limit imports of parts because of the unavailability of foreign currency. The cumulative result of all these factors was a reduction in the production of motor vehicles.

The creation of the Ministry of Heavy Industry and the subsequent decision to produce commercial vehicles such as trucks, buses, pickup trucks, and minibuses to meet domestic needs were important events during this time. In this regard, extensive research and studies relating to the design of commercial vehicles and the improvement of domestically produced parts were undertaken, and from at that point, the idea of producing domestically designed and manufactured automobiles emerged. Simultaneously, due to shortages of automobiles that resulted from price controls, a black market for cars surfaced in Iran. The state management of the auto companies, which did not have development plans, did not have the ability to produce parts domestically, could not increase the productivity of the existing auto factories, and eventually, by increasing the bureaucratic red tape, compounded the difficulties. Finally, in

1985, a plant to produce transmission systems was constructed, which, in addition to the production of transmissions for a new generation of automobiles, also produced motorcycles (Manteghi 2011).

The Years of Lost Opportunities (1995–1999)

After the end of the Iran-Iraq war in 1988 and during the implementation of the First Economic, Social and Cultural Development Plan of the Islamic Republic (1989–1993), the government adopted a policy of structural adjustment intended to bring fundamental changes in the national economy. During this time, the import of automobiles to Iran increased drastically due to the pent-up demand for goods following 8 years of war, and a rise in Iranian oil revenues stemming from a reduction in the export of oil from Iraq as a result of the economic sanctions against that country, which had boosted the price of petroleum,. The number of imported cars increased from 103 units in 1988 to 500 units in 1989, then to 7,296 units in 1990. Finally, the number of imported cars in 1992 rose to 33,950 units, which was higher than the number of units produced domestically. Furthermore, during this time certain policies related to the automobile industry were adopted, which are described below:

1. In 1998, the Ministry of Heavy Industry adopted the policy of acquiring self-sufficiency in auto production, and to achieve this goal the Sazeh Gostar Company was established to produce automobile parts.
2. In 1999, the Center for Automobile Policy Making was established in the Ministry of Heavy Industry, and to remedy the lack of access to the suitable molds for design of new metal sheets for manufacturing auto bodies, formed Iran Heavy Dies Manufacturing Company.
3. In 1999, after 5 years of prohibiting the import of cars, the government relaxed the prohibition somewhat under certain conditions; however, the prohibition was imposed again a year later.
One of the reasons for the prohibition of the import of automobiles was inadequate postsales services. The repair costs of imported lemons were very high, and the quality of services to the customers was poor. In 1991, after the passage of a law, the import of foreign manufactured cars was allowed only on the condition that dealerships of foreign manufacturers had a local presence in Iran. At that time, only four companies met this requirement.
4. In 1992, an important step in encouraging the automobile industry to produce auto parts in Iran was the passage of the

Automobile Law. This law created a framework for the adoption of stable policies for development of automobile technology in the county. The foundation of this law was to achieve the self-sufficiency of the automobile industry in the country. This law emphasized increased automobile production, recommended higher output of domestically produced auto parts, required improved quality of the cars, and mandated provision of after-sales services in Iran. Moreover, this law emphasized complete indigenous design and manufacturing of automobiles, where all the major auto parts were to be produced in Iran. This law levied a 220 percent tariff on imported cars, which in practice led to no import of automobiles to the country.

5. In 1993, Mega Motor Company was established for the purpose of achieving self-sufficiency, and for the production of engines, transmissions, and axles. This company relied on available resources in the Saipa Company to produce these items. Furthermore, the Center for Research and Innovation of Automobile Industry was also established as part of the Mega Motor Company.

During this period, Iranian automobile companies, by entering into many agreements with transnational automobile companies, produced a variety, albeit a limited number of automobiles (Manteghi 2011).

Actors in the Iranian Automobile Industry

In this section, we introduce the major players in the automobile industry, which consist of domestic car and auto-parts producers, importers of automobiles, and the executive as well as legislative branches of government.

Domestic Automobile Producers

We introduce major Iranian vehicle producers and the total output of each company in 2010 in Table 10.1 (Iranian Automaker Association 2012).

Based on the data in Table 10.1, it is clear that 95 percent of the total output of motor vehicles in Iran is produced by only two firms: Iran Khodro Company, including Iran Khodro Diesel; and Saipa Company, including its units such as Saipa Diesel, Pars Khodro, and the Zamyad Company. This implies a two-firm concentration ratio of 95 percent.

Until 2010, the Industrial Development and Renovation Organization of Iran (IDRO), a state-owned organization, was the main shareholder

Table 10.1 Iranian Automobile Firms and Their Output, 2010*

Company	Production of Automobiles Including Vans	Company	Production of Commercial Motor Vehicles*
Iran Khodro	751,624	Bahman Diesel	14,468
Saipa	383,147	Iran Khodro Diesel	10,380
Pars Khodro	304,448	Saipa Diesel	8,356
Zamyad	610,80	Carizan Khodro	2,995
Bahman Industrial Group	395,65	Aras Khodro Diesel (AMICO)	2,149
Modiran Vehicle Manufacturing Co.	7,144	Oghab Afshan	1,249
Rayen khodro	4,500	Rakhsh Khodro Diesel	1,234
Sanabad Khodro	538	Zamyad	1,102
Bam Khodro	311	Run Iran	462
Diar Automobile Co.	238	Pishro Diesel	142
Morattab Khodro	180	Horand Khodro Diesel	113

*Firms with fewer than 113 units of output are not listed.
Source: Manteghi 2011.

of these companies. It should be noted that these state-owned companies were managed as private for-profit enterprises; however, until 2010, the Iranian government appointed the corporate executives of these enterprises. In a recent privatization of state-owned enterprises, Iran Khodro Workers' Cooperative became the main shareholder of Iran Khodro, Saipa Investment Company acquired controlling interest of Saipa, and IDRO lost control of the companies.

Iran Khodro as the leader and Saipa as the follower companies have devised policies for Iran's auto industry in recent years. As such, in the next section we only trace the evolution of these two companies.

Industrial Development and Renovation Organization of Iran

The Industrial Development and Renovation Organization of Iran was set up in 1967. The need for the presence of a powerful, all-embracing organization for industrial development led to the establishment of this organization.

During IDRO's first decade of operations, its mission was the "completion of a chain of industrialization, with strategy of import substitution." Its current mission, however, is "expansion and renovation of Iranian industries with the aim of making these industries globally competitive and securing the interests of the stakeholders" (IDRO 2012).

At the time of this writing in January 2012, IDRO owned about 20 percent of the stocks of both Iran Khodro and Saipa. The management of IDRO's interests in the motor-vehicle industry is a responsibility of the Deputy Director for Renovation and Revenue Generation. The goals of this organization in terms of the motor vehicle industry are to increase efficiency, reduce costs, improve quality, provide better after-sales service, maintain the state-of-the-art quality of the products, and make the products globally competitive (IDRO 2012).

Auto-Parts Manufacturers

In 2011, there were two thousand motor-vehicle parts manufacturers in the country. We may classify these manufacturers into three categories according to the quality of the products they produce: A, B, and C. There exist thirteen hundred active manufacturers in categories A and B. However, due to the low quality of products, the producers in category C in practice receive no orders and are gradually withering away (Iranian Auto Parts Manufacturing Association 2012). Moreover, approximately 70 percent of the parts of domestically assembled vehicles are manufactured in Iran, and the rest are imported. Some of the parts of vehicles assembled in Iran, which are intended to be produced inside the country, are manufactured in countries like China, Egypt, Ethiopia, and Vietnam.

There are two reasons for the importation of some parts although the technological capabilities for producing them exist domestically. First, many domestic car manufacturers do not place adequate trust in domestically produced parts. Second, the auto-parts producers, because of low demand for some parts, face diseconomies of scale, that is, high unit cost due to inadequate scale of production (Soltan Zadeh 2012). The value of the output of auto parts produced in Iran was $700 million in 2010.

The major motor-vehicle producers in Iran purchase the parts they use in assembling vehicles through their subsidiaries. Specifically, Iran Khodro purchases parts through its subsidiary, Shopco Company, and Saipa Company procures parts via its subsidiary, Sazehgostar Company.

The Importers

Currently, four requirements for the importation of motor vehicles consist of acquiring permits from (1) Organization of Improving Efficiency of Fuel Consumption; (2) Environmental Protection Organization; (3) Organization of Standards and Industrial Research; and (4) a service center for the provision of after-sales services in Iran (Ministry of Industry, Mines and Commerce 2012).

Some of the well-known brands that have supplied motor vehicles in Iran include Mercedes Benz, BMV, Toyota, Hyundai, Nissan, Lexus, Kia, Peugeot-Citroen, and Alpha Romeo.

The Government

The most important impact that the government has had on the motor-vehicle industry in Iran is through the levying of tariffs on the importation of motor vehicles into the country. Previously, the Ministries of Industry, Mines and Commerce as well as Economic Affairs and Finance were responsible for determining the tariff rate. The rate of tariff for the importation of motor vehicles was 90 percent in 2011. To join the World Trade Organization (WTO), the government must lower the tariff rate for importation of vehicles. However, aspects of the Iranian motor-vehicle tariff system create a shadow of doubt about achieving this goal. First, there exists no long-term tariff policy in the country, and second, the rate is determined according to the required financial needs of the government during the annual government budget-negotiation process between the executive and legislative branches of the government.

Moreover, the government, through IDRO, one of its industrial-development arms, has promoted the technological capabilities of the motor-vehicle industry by assisting in investment in technology transfers. For example, collaboration among IDRO with a 40 percent interest; the Kartak System Company, a private motor vehicle enterprise with a 40 percent interest; and the Italian Pistal Racing Company with a 20 percent interest, resulted in the establishment of a pistons production unit at the Caspian Industrial Estate in Qazvin, a city near Tehran. The unit became operational in August 2011 with the capacity to manufacture 1.3 million pistons for gasoline engines and 700,000 pistons for diesel engines.

In the production of these pistons, advanced automatic machinery and robotic systems in casting, cutting, and feeding operations are used. Moreover, automatic monitoring and controlling units, as well as advanced metallurgical and metrological laboratories, guarantee the manufacturing of high-quality products. The Pistal Racing Company has acquired these advanced technologies by cooperating with leading

European motor-vehicle companies such as Renault, Peugeot, Ferrari, and Jaguar (IDRO 2011). The Iranian government is also contributing to the growth and technological development of the motor-vehicle industry in another way. The newly formed Ministry of Industry, Mines and Commerce, which was created by merging the Ministry of Industry and Mines with the Ministry of Commerce, intends to set policies to improve the quality of motor vehicles produced in Iran so that at least 30 percent of these motor vehicles can be exported.

The Legislating of Motor Vehicle Laws

The Ministry of Industry, Mines and Commerce is responsible for informing the motor-vehicle companies about the new laws relating to the motor-vehicle industry. The most important laws relating to this industry in Iran consist of those pertaining to:

- Standards for production, safety, and operations (some of the standards are set by the Institute of Standards and Industrial Research);
- Environmental protection standards;
- Rules and regulations concerning protection of the consumers of motor vehicles;
- Fuel efficiency standards for the vehicles sold in the domestic market.

The Path for Globalization of the Iranian Motor Vehicle Industry (1995–2011)

In this section, we discuss the development of the industry along the path of indigenous technological-capacity building and the globalization of Iranian manufactured motor vehicles that took place between 1995 and 2011.

The First Step: Becoming Automobile Builders (1995–2001)

Several years before this period, the government encouraged banks, all of which were government owned at that time, to allocate long-term credit to large industrial enterprises, including the auto industry, by guaranteeing the loans. However, the government had no policies concerning the import of raw materials, intermediate inputs, parts, and manufactured products to the country. No consideration was given to the potential demand of these imported goods in the domestic markets, and only the profit motives of importers determined what was imported. Specifically, due to a lack of economic calculations and the unrealistic optimism of importers that they could sell the imported intermediate goods in the Iranian markets, massive imports in excess of the required inputs to produce the output of the domestic industries led to trade deficits and the inordinate outpayment of foreign exchange reserves.

The politicians and industrial policy-making authorities of the country took measures to protect industries in general in response to the balance-of-payments difficulties and the challenges facing industry, and the motor-vehicle industry in particular. The law prohibiting the import of motor vehicles was such a measure. The prevailing economic conditions in the country at the time required the policy makers to adopt policies based on the domestic capabilities of the country.

Iran Khodro, as the largest auto-manufacturing company in the country, and inspired by the transnational auto companies, adopted the first harmonious strategy for becoming an own brand manufacturing (OBM) enterprise during 1995–2001. The goals of this strategy included moves toward becoming OBM, increased variety of models and total output, increased domestic production of parts, increased capacity to produce output, and expansion of the subsidiary companies.[1]

During this time, the Samand, the first domestic brand automobile from design to mass production, was produced.[2] The design of the Samand started in the second half of 1995, and 26 units of this car were produced as test models by 2000. Commercial production of the Samand began in the third quarter of 2001 (Manteghi 2011).

Accordingly, Iran Khodro took the first step to becoming an OBM auto firm. It should be noted that often the latecomer auto firms use the platform of a well-established car manufacturer before a complete design of the car gradually takes place. The platform for the Peugeot 405 was used for production of the Samand.

Due to space limitations, we are not able to provide the time-series data for the annual output of the Samand brand. Suffice to say that in 2001, the company produced 1,345 Samands, and by 2010, this number increased to 133,091 units (Association of Auto-makers of Iran 2011).

The activities of Saipa Company during this period include:

1. Implementation of extensive activities for domesticating the production of auto parts in 1998;

2. Production of the Pride brand with the use of 81 percent of the value of the car from domestically produced parts and the manufacturing of 79 percent of the value of Nissan cars by domestically manufactured parts in 1999;
3. Signing of a technology-transfer agreement with the French company Citroen and the production of a car called the Zanita in 1999;
4. Production of Pride vans during 1999 and 2001;
5. Improvement in the appearance of the Pride, and the production of fuel injection for the car in 1999;
6. Acquisition of self-sufficiency in the production of Nissan cars; and
7. Completion of the assembly line for Karavan vans, production of test units in 1999, and commercial production of the vans in 2001[3] (Saipa, 2012).

In designing Karavan vans, Saipa did not follow the common pattern of using an existing platform, but rather began the design of these vans by collaborating with small Italian automobile design firms. Due to inadequate quality control, even though the numbers sold increased for the first few years after its introduction, the number of Karavan vans sold never increased beyond nine hundred units. After the peak year of 2005, sales dropped, and eventually the company ceased production of Karavan vans in 2010.

Toward Globalization (2001–2011)

In this section, we will discuss the developments in Iran Khodro and Saipa during this period.

Iran Khodro

Iran Khodro implemented a strategy to become a global company and close the technological gap between itself and major transnational motor-vehicle manufacturers during the first decade of the twenty-first century.

In this period, the company made use of planning for exports to international markets; methods such as the export of passenger cars, commercial vehicles, rail vehicles, and parts; and the establishment of manufacturing facilities for the production of national brands. The strategy to increase exports of Iran Khodro products was in harmony with the 20-Year Technological Vision of the country and its orientation to promote exports.[4]

For entry into the international competitive arena, conditions such as the design of an effective strategy, the reduction of costs, the establishment of internationally recognized quality standards, the use of modern marketing techniques, the supply of a variety of differentiated products and services, the provision of global security, and product and process innovations had to be realized. It was necessary for Iran Khodro to combine two strategies for becoming globally competitive, that is, inward-looking and outward-looking approaches. The combined strategy consisted of absorbing the most advanced automotive technologies available from abroad for use in the production of motor vehicles domestically for the export markets.

The outward-looking strategy involved the transfer of automotive technology from transnational automakers, the improvement of transferred automotive technologies, and the export of the manufactured motor vehicles and parts to the original auto manufacturers. The inward-looking strategy required the establishment of production sites at the targeted markets, the export of motor vehicles and parts to those markets, and the establishment of service centers and provision of after-sales services.

Some of the activities relating to becoming inward-looking and globally competitive include:

1. Cooperating strategically with the leading global automotive companies, for example, a joint venture with French Renault to jointly produce the Renault L90;
2. Placing parts makers into the international network of prominent automotive parts producers, and exporting the parts;
3. Designing automobiles and parts in collaboration with prominent automotive producers, with the aim of transferring advanced technologies from other countries, especially regarding soft know-how such as marketing and after-sales services, and the export of motor vehicles using the network of prominent manufacturers. Specific examples include an agreement with Peugeot for the export of the Peugeot 206 with a luggage compartment, with Renault for the export of the pickup PU90, and with a Chinese company, Chery Automobile, for the export of brand S21 automobiles.

In spite of the widely held philosophy that governments should not become involved in management of the private companies, the legislative branch of the Iranian government passed a law that nullified an agreement with Chery Automobile. The reason for this law was the requirement of establishing service centers in the country of sales, which Chery Automobile could not fulfill.

Some of the activities regarding the outward-looking strategy for becoming globally competitive include:

1. Creating a brand for Iran Khodro, and introducing it to international markets;
2. Using platforms and technological capabilities of leading automotive companies in the framework of strategic cooperation;
3. Producing the Samand according to international standards for the export markets.
4. Establishing a production line by creating joint ventures in the target markets.

Moreover, Iran Khodro, for the purpose of becoming a global player, transferred technologies to other countries. These transfers for a period of 5 years (2001–2006) consisted of the following activities.

Important events took place from 2001 to 2006, which is considered a period of relative maturity of the industry in terms of research and development (R&D), and the beginning of a global presence on the part of the Iranian automotive companies. These notable events consisted of acquiring the prevailing production methods in the global auto industry with the applications of modern technologies on the production lines, improving the quality of products, supplying some products that meet international standards, and entering into international markets. The international markets that Iran Khodro has entered include Russia, Ukraine, Azerbaijan, Belarus, Algeria, Tunisia, Senegal, and Venezuela.

- These market entries were made possible through cooperative arrangements with transnational automotive companies. For example, a $375 million joint investment involved the IDRO (23 percent), the Renault-Nissan Company (51 percent), Iran Khodro (13 percent), and Saipa (13 percent). In another strategic alliance between Iran Khodro and Peugeot Company for the design of the Peugeot 206 SD brand, Iran Khodro contributed 90 percent of the 100 million Euro investment.
- The technological gains during this period consisted of the design of a propulsion system, the ability to design a motor vehicle based on native technologies, the design of complete lines of production, the ability to manufacture motor vehicles according to standards

of the export markets, and the acquisition of marketing and service skills. Finally, Iran Khodro Company learned how to design and produce motor vehicles by cooperating with transnational auto firms for export markets.

2006–2011, The Years of Becoming Globally Competitive

The main strategy during this time was completion of the Iran Khodro globalization process. The most important goals of this period consist of the following:

1. Developing the ability to design suitable vehicles for target markets;
2. Developing the ability to design a family of propulsion systems based on three types of fuel: Gasoline, diesel, and natural gas;
3. Developing the ability to design different components of an automobile;
4. Developing the ability to form platforms independently and collaboratively;
5. Establishing a global reputation as an innovative automotive company that has developed engines that use natural gas as fuel;
6. Becoming competitive in all products in the target markets;
7. Becoming the center for R&D in the electronic components of automobiles.

The technological strategies of Iran Khodro to achieve the above capabilities include:

1. Developing an integrative technological system for acquiring maximum profitability;
2. Developing international cooperation for the acquisition of soft technologies, that is, marketing and providing customer services for the design of vehicles and the components of vehicles;
3. Utilizing the existing research network for the creation of endogenous technologies and their constituent parts for motor vehicles;
4. Engaging in research on subjects such as automotive electronics, natural gas-based engines, hybrid vehicles, platforms, paints, and polymers;
5. Establishing first-rate producers through the adoption of information technologies for competitive prices and the transition to becoming an exporter of parts;

6. Securing a position in the international network of motor-vehicle producers by using information technologies to become price competitive;
7. Institutionalizing the idea of the significance of technological innovation at all levels of the organization.

Some of the achievements of the Saipa Company during this period are listed below.

1. Producing a limited number of the Pride 141 as test models, and mass production of this product in 2002;
2. Producing the Pride with a fuel injection system for the first time in 2003, and mass producing this engine a year later;
3. Producing minibuses and pickup trucks with dual fuel-burning (gasoline and natural gas) capability in 2003, and setting up a production line for the production of dual fuel-use capability in the country in 2004;
4. Establishing a production line for the Pride in Venezuela in 2005, and starting production of L90 cars in cooperation with the Renault Company in 2006;
5. Establishing assembly lines for the production of a number of models of the Pride in Syria in 2006. The total investment in this project was $20 million with an estimated 100 percent return on investment in 3 years;
6. Introducing a new car named the Taipa in 2007;
7. Introducing a new engine that was designed and manufactured by the Saipa Company in 2008;
8. Introducing the Kaveh truck, the first truck manufactured by Saipa in 2010 (Saipa 2012).

Additional Comments

Share of Domestic Car Producers in the Domestic Market

The average production of Iranian automakers was one million units over the last five years. The maximum number of imported automobile was five thousand units per year from 2006–2011. The Iranian automakers' share of the Iranian auto market is 95 percent. These numbers indicate that with a big wall of protective tariffs, Iranian automakers face no threat and no competitive pressures from transnational automotive companies. However, transnational automakers control the market for luxury cars, which constitutes only 5 percent of the Iranian auto market.

It is instructive to mention that Iran ranked as the sixteenth-highest producer of automobiles in the world in 2005. The ranking increased to the tenth highest in 2009; however, it dropped to the twelfth highest in 2010 (The International Organization of Motor Vehicle Manufacturer (OICA) 2010).

Employment

The number of workers employed by the Iranian auto industry and related businesses is estimated to be in the range of 400,000 to 500,000. The estimate is based on employment data that was published by the Statistical Center of Iran in 2007.

International Cooperation

In this section, we provide a list of major Iranian automotive companies and their international partners. We first provide a list of automobile firms, and then introduce companies that produce commercial vehicles.

Among the leading automobile firms, we cite Iran Khodro, which cooperates with Peugeot, Renault, and Suzuki; Saipa, which works with Kia and Citreon; Pars Khodro, which has agreements with Kia, Renault, and Nissan; Zamyad, which collaborates with Nissan and IVECO (Italy); Bahman Industrial Group, which has engagements with Mazda, Mitsubishi, ZX, and FAW, the latter two of China. Modiran Vehicle Manufacturing Company and Sanabad Khodro both cooperate with Chery, a Chinese firm; Rayen Khodro works with Hyundai; Bam Khodro has agreements with Volkswagen, Hafeiauto, and Liftan Motor, the latter two firms being Chinese automakers; and finally, Morattab Khodro has a working relationship with Musso, a Korean firm.

The Iranian commercial vehicle producers that are involved with foreign automakers are Bahman Diesel, which works with ISUZU and FAW; and Iran Khodro Diesel, which cooperates with Daimler Benz, Hyundai, GAZ (Russian), and Sinotruck (Hong Kong). Furthermore, Saipa Diesel cooperates with Renault and Volvo; Oghab Afshan deals with Scania (Sweden); Rakhsh Khodro Diesel has agreements with Kamaz (Russia), JAC, and Jinbei, the latter two Chinese firms; Zamyad works with IVECO; Run Iran collaborates with Volvo; Horand Khodro Diesel has a working relationship with Dongfeng Motor (China); Yavaran

Khodro works with MAN (Germany); and Arya Diesel and Shahab Khodro both cooperate with Renault.

Iranian Laws to Develop the Automotive Industry

In the winter of 2009, the Council for Automotive Policy Making, which is an administrative body of the former Ministry of Industry and Mines adopted Aims and Policies Relating to Development of Automotive Industry in Accord with Vision 1404. The policies articulated in this document could be considered the government's planning strategies for the development of the automotive industry. Among the main objectives, we cite encouraging, close interfirm collaboration for the purchase and transfer of automotive technologies, the establishing of close working relationships between firms and universities, the setting up of science and technology parks, and support for consulting and engineering service firms.

The Vision of the Industry over the Next Two Decades (Vision 1404)

The 20-year vision for the auto industry is defined as "achievement of the first-rank producer of motor vehicles in the region,[5] the fifth-highest producer in Asia, and eleventh-largest producer in the world by becoming competitive through the development of technology." According to global automobile production statistics, Iran has achieved the first two goals, and has almost achieved the third one by ranking twelfth globally in 2010 (OICA, 2010).

The objectives in achieving these goals include the establishment of production plants for making purely Iranian brand names, joint Iran-foreign brand names, or global brand names in the region with an emphasis on production of output for export markets; and the establishment of production units for the manufacturing of automotive parts with prominent domestic or international trademarks in the region with an emphasis on competitive advantage. Furthermore, other objectives are to attract domestic and foreign capital in the form of direct investment or joint ventures, with the aim of acquiring modern technologies and expanding exports, and establishing research, training, and engineering services centers in the region.

The Goals of the Automotive Industry in Vision 1404

The goals of the Iranian automotive industry as defined by the Council may be divided into two parts. First, the goals pertaining to the light vehicles, that is, automobiles and pickup trucks consist of the following items:

1. Annual production of three million units, with two million units for the domestic market;
2. Production of at least 50 percent of light vehicles under domestic brand names;
3. Production of parts with a value of at least $25 billion for domestic car makers;
4. Annual export of $6 billion of parts for the independent foreign markets, that is, assembly lines and spare parts.

The goals for the commercial producers consist of the production of at least 120,000 units, with 90,000 units for the domestic market. Moreover, the automakers aim to increase the value-added of the automotive industry to at least 45 percent of the domestic gross national product of Iran. Finally, they plan to increase the value-added of the automotive industry to 19 percent of the value-added of the total industrial sector of the economy.

Recommendations for the Necessary Paths for Achieving the Goals

Achieving the above-cited goals requires the purposeful, harmonious, and long-term support of the government for the process of industrial development of the country through means such as subsidized bank credit allocation, subsidies for total or, at least partial, costs of R&D of new designs with modern technologies, tax write-offs of the expenses of the procurement of software and hardware of assembly lines using new technologies, and the provision of special access to the use and preferential treatment of industrial enterprises in obtaining foreign exchange. The government could create the following developmental paths for achieving the goals and for the industrial development of the country.

Completion of the privatization of state-owned enterprises (SOEs) in accord with Article 44 of the Iranian Constitution, which prohibits the government from increasing its holding of enterprises and by implication, encourages privatization.

Increase efficiency by identifying and remedying factors that inhibit efficiency, specifically by:

1. Creation of a stable legal environment based on transparency and known international standards;

2. Establishment of legal reforms and a stable monetary policy;
3. Establishment of legal reforms relating to property rights, anti-dumping, labor, social security, and corporate governance, including laws relating to holding companies, mergers and acquisition, dissolution of companies, and commerce;
4. Avoidance of interventions in the markets, especially by imposing price controls, which always leads to misallocation of resources;
5. Facilitation of the processes of foreign direct investment as well as joint investment with foreign entities that are aimed at the transfer of modern technologies and the use of their sales networks for exports;
6. Participation in the regional and global agreements.

The Automotive Industrial Policies of the Ministry of Industry, Mine and Commerce

Outside of the Law of Motor Vehicle that was passed in 1992, which was discussed above, the Ministry did not take any policy measures concerning the Iranian automotive industry until 2010. In 2010, this ministry adopted the standard industrial policies, as noted above, that are to be implemented by 2023.

In the spirit of governmental support of the industry, we recommend government support of the automotive companies in achieving the goals by raising the environmental and safety standards of motor vehicles to meet European standards, and by gradually ending the production of motor vehicles by SKD[6] and FCKD[7] methods.

The Responsibilities of the Automakers

The Council for Automotive Policy Making, in its Aims and Policies Relating to Development of the Automotive Industry in Accord with Vision 1404, enumerated the responsibilities of the automakers as follows.

1. Accept the industrial goals and policies of the country;
2. Accept the directives of the Ministry of Industry, Mines and Commerce, by implementing the safety and operational standards of production processes, that is, establishing conformity of production and quality-control systems;
3. Implement environmental standards;
4. Produce and distribute regular reports of the facts of operations and financial conditions of the firm based on well-defined indicators.

5. Implement rules and regulations relating to the protection of consumers;
6. Produce motor vehicles according to the fuel efficiency standards for vehicles sold in the domestic market (Ministry of Industry, Mines and Commerce, 2012).

The Main Challenges Faced by Automakers

We classify the challenges that the Iranian automakers face into internal and external categories.

Internal Challenges

We recognize that the following assessment might be subjective. However, it is based on the author's experiences as the chief executive officer of the leading Iranian automotive company, Iran Khodro.

We believe the Iranian auto industry suffers from a weak corporate governance structure, a low quality of some of the automobiles it produces, a lack of variety of automobiles supplied to the market, and a weakness in technical knowledge. Moreover, we believe that due to diseconomies of scale, investment in R&D is not cost effective. Other difficulties facing the industry include the high cost of production, the limited size of the domestic market and difficulties of exporting the products, the weakness of processes and systems, a lack of expertise in the organizations, low productivity of labor and capital, and finally weaknesses in marketing and the technologies of enterprises.

External Challenges

The external challenges facing the industry may be classified into three categories: challenges emanating from the economic and industrial substructure, challenges emerging from the legal substructure, and finally, challenges resulting from the political and cultural substructure. We list these challenges for each category.

Economic and Industrial Substructure

The difficulties under this category consist of a shortage of liquidity and inadequate financing options; the high risks of capital investment; high inflation; a weak domestic currency; inadequate foreign investment in Iran; the passage of commercial laws, rules, and regulations that are heavily influenced by import-substitution policies; a lack of adequate attention to modern, advanced technologies; inward-looking business and economic strategies; imbalances in value-added across industrial sectors of the economy; the dominance of the government in the economy; a and lack of coordinated macroeconomic policies.

Legal Substructure
The legal structure of Iran does not allow for freedom of action of public-owned enterprises. State-owned companies face legal difficulties pertaining to the labor laws, tax laws, foreign investment laws, private property rights, and intellectual property laws.

Political and Cultural Substructure The challenges in this category include strong anticapitalist sentiments among the population, the lack of desire to work hard, the absence of an intense desire to achieve, and inadequate discipline in the workplace. Moreover, in spite of successes in some technological fields in Iran, there exist some deficiencies in the national system of innovation[8] (Manteghi 2011).

Summary and Conclusion

The Iranian automotive industry has been profoundly influenced by socioeconomic and political events of the last half century in the country. These events have had a drastic retarding effect on the nature and rate of growth of the industry, to the extent that more than 70 percent of the life of the firms in the industry were merely assembling parts for production of motor vehicles.

In the second half of the first decade of the twenty-first century, the industry has had a double-digit annual growth rate, is in good condition from a production point of view, and more than 70 percent of the vehicles it has produced and that are currently on the road were more than 20 years old in 2012.

To a certain extent, the acquiring of the technological knowledge, in cooperation with certain prominent world automakers, is underway and has been manifested in domestically produced vehicles. However, protective tariffs have granted the industry, in fact, only two leading firms, monopoly status. Moreover, due to the traditional, commercial (in contrast to industrial and innovative) views of many auto executives, their inability to affect innovation in their firms, and limited government involvement in promoting the industry's investment in R&D, the development of the automotive industry has occurred at less than the expected world standards. This has led to an inadequate growth of the industry in global competition. The gap between the Iranian automotive industry and the leading international automobile industries on the one hand, and Iran's desire to join the WTO, which necessitates a large reduction or elimination of tariffs on the other hand, requires accelerated growth in quality and innovation in the design of automobiles so that Iran can become successful in competing with transnational automakers in the coming years.

Achieving this goal necessitates a precise, sensible policy that should be formulated by the Ministry of Industry, Mines and Commerce as well as the IDRO. Moreover, reaching this goal requires careful oversight of the implementation of automakers' strategies, the creation of suitable components, a concentration on the development of domestic brands, a close working relationship with the universities, a presence in international export markets and an increase in the share of exports, and the qualitative growth of domestic auto-parts makers. Of course, the industrial achievements of the last ten years in Iran, especially at Iran Khodro, show that this goal is achievable. Recently, the strategy for Iran's automotive industry was formulated by IDRO, although this strategy is not supplemented by specific operational measures and objectives. It is more a statement of desires and hopes; however, concentrating on it and reviewing it annually could result in a very effective plan for operations for the development of Iran's automobile industry.

Notes

1. For a discussion of the Own Brand manufacturing company, see the introduction of this volume.
2. We refer the interested reader who wishes to see a photograph of this and other brands of Iran Khodro to http://www.ikco.com/En.
3. To see photographs of Saipa-manufactured vehicles, visit http://www.saipacorp.com.
4. For a discussion of the 20-Year Technological Vision of the country, see chapter 4 of this book on the national system of innovation.
5. The Middle East, and Central/Western Asia.
6. Semi Knocked Down (SKD): In this production method, an incomplete kit (containing semimanufactured parts that are needed to assemble a product, e.g., an unpainted complete body, engine, gearbox, etc.) that is typically manufactured in one country or region, is exported to another country or region for final assembly.
7. Full CKD, a kit containing the parts for assembly of a product.
8. We refer the reader to chapter 4 on the national system of innovation in this book for detailed discussions of this topic.

References

Association of Iranian Automakers. 2012. "Reports and Statistics" (in Farsi). Ivma.ir. Accessed January 20, 2012. http://ivma.ir.
Industrial Development and Renovation Organization of Iran (IDRO). 2012. "Mission and Goal" (in Farsi). Idro.org. Accessed January 20, 2012. http://www.idro.org.
———. 2011. "Through Cooperation of IDRO with the Private Sector Car Pistons Production Plan to Come on Stream. Idro.org. August 1. http://www.irro.org.
International Organization of Motor Vehicle Manufacturer (OICA). 2010. "2010 Production Statistics." Oica.net. Accessed February 14, 2012. http://oica.net/category/production-statistics.
Iranian Auto Parts Manufacturing Association. "About Us" (in Farsi). Iapma.ir. Accessed January 30, 2012. http://www.iapma.ir.
Manteghi, M. 2011. *Towards Globalization* (in Farsi). Tehran: Ideh Nu Publishing.
Ministry of Industry, Mines and Commerce. 2009. "Ahdafvasyasathaitose a sanaatkhodrudarofogh 1404" (Goals and policies of development of automotive industry in year 1404 horizon). Mim.gov.ir. Accessed February 12, 2012. http://www.mim.gov.ir.
Parsa, H. M. 2009. *The First Complete Encyclopedia of Automobiles in Iran* (in Farsi). Tehran: Amir Ali Publishing.
Saipa Automotive Manufacturing Group. "About Us" (in Farsi). Saipacorp.com. Accessed February 15, 2012. http://www.saipacorp.com.
Soltan Zadeh, M. 2012. "Small Share of Domestic Auto Parts in Assembled Automobiles" (in Farsi). Iapma.ir. Accessed February 15, 2012. http://www.iapma.ir/fa/index.php?option=comcontent&view=article&id=2160:1390-10-14-11-38-31&catid=1:latest-news&Itemid=102.

CHAPTER 11

Conclusion

Sepehr Ghazinoory and Abdol S. Soofi

A comparative study of Iran with other countries shows that Iran has a set of unique features. Accordingly, understanding the developmental processes of science and technology in Iran requires informed insights into the historical, geographical, cultural, political, and social characteristics of the country.

First, the Islamic Revolution in 1979, and the subsequent establishment of a theocratic state have never occurred elsewhere in the recent history of mankind. Second, a variety of nationalities constituting the Iranian population adhere to the Shia' faith, which unifies the Iranian people and gives the country a sense of national identity. Third, the natural resources, particularly the rich hydrocarbon deposits, have turned the country into a leading source of petroleum, natural gas, and other mineral resources such as copper. Fourth, the geopolitical situation of Iran, which is located in the center of the main source of energy supplies to the rest of the world—the Persian Gulf and central Asia with the latter's vast untapped energy resources—is another differentiating feature of the country. Finally, the rapid growth of young, well-educated human resources is another distinguishing attribute of this ancient land.

A cursory review of the history of Iran (Persia) shows that the country has faced numerous foreign invasions, internal strife, and power struggles. The Islamic Revolution of 1979, by a reliance on the Islamic faith and a mass-based ideology that is extremely sensitive to the needs of the underprivileged masses, aimed to eliminate political instability in the country, and tried to establish political unanimity in the framework of being responsive to the needs of the Iranian peoples, a unanimity that is a prerequisite for building a lasting civilization. Nevertheless, political instability persists, and the political discord has created difficulties for the full development of science and technology in the country.

Almost all contributors to this volume point to a similar set of difficulties in achieving the technological goals, issues such as international sanctions, the brain drain, lack of competition in the industries, inadequate infrastructure, the overall political frictions, and macroeconomic problems.

These problems do exist; however, the authors offer many examples that show that because of these lingering difficulties, reaching an innovative technological system is rather challenging for firms in many industries in the country. However, one of the counterexamples for successful innovation is the technological advances of the Iranian defense industry.

The Iranian defense industry has faced strict sanctions since the triumph of the Revolution in 1979. This industry, consisting of mostly government-owned enterprises, has a competitive disadvantage relative to civilian industry in the sense that it has the least capability to attract the most talented human resources. Massive bureaucracy and lack of competition are also the main characteristics of the sector. However, based on the findings of Fartookzadeh and Vaziri (2008), the industry has accomplished notable innovations, which has increased the strength of the Iranian national defense considerably.[1]

Now, a number of important questions must be answered. For example, "Why was the Iranian defense industry more innovative compared to the civilian industries in spite of the difficulties stipulated above?" Or "Why has the Iranian oil industry failed to innovate and repeatedly had to purchase the same license to use a technology from foreign suppliers?" Furthermore, "Why is the oil industry incapable of developing the same techniques at home, but the firms in the Iranian defense sector, in spite of the tough sanctions and without foreign licensing, have developed many new processes and product innovations?

Answering these questions requires extensive study; however, one could speculate on some of the underlying causes of this unbalanced technological development. Below, we list some of these factors that may have contributed to the phenomenon.

First, science and technology policy planning in the defense sector is very focused on technological development. This focus is a matter of necessity because most military equipment involves high technology.

Second, the monopsonistic (single-buyer) nature of the industry plays an important role. Facing military threats from the most powerful militaries of the world, the Iranian armed forces have high demand for technologically advanced military hardware. On the supply side, two hundred companies produce the required goods and services for the armed forces. In most cases, the military has to procure the required equipment from a few, in many cases from single, suppliers, and does not have much freedom of choice in selecting among the many competitors. Accordingly, the procurements of the military hardware require the military's use of methods such as feedback, exertion of political pressure, and joint learning activities to improve the technical know-how of the producers. Many institutions such as the Ministry of Defense and its research and development (R&D) organizations have played the role of technology broker.

Third, the relative technological success of the Iranian defense industry is due in large part to its insulation from domestic political pressures. The political insulation emanates from the defense establishment being under the direct control of the Supreme Leader and not answerable to the Office of the President, which must engage in a balancing act of coordinating among the interests of different constituencies.

Finally, the military's contacts with the civilian research institutes and its ability to use the best technical capabilities of the latter may be considered as important contributing factors for the differential performance of the defense sector.

Whatever the reasons for the successes of the Iranian defense industry, they indicate that all of the difficulties facing technological development in the country that were specified in the chapters of this volume could be overcome if the correct policies of the government are combined with the efforts of researchers and technologists. Such a combination could lead to the innovation stage, even in the face of extensive international sanctions.

One should not forget the proverb, "Necessity is the mother of invention."

Note

1. Due to space limitations, we are not able to enumerate these achievements; however, some of these technological advances are discussed in various chapters of the book.

References

Fartookzadeh, H., and J. Vaziri. 2008. "Creating Defense Competency in Fourth Wave: A Study of Implementing Networking Approach in Knowledge-Based

Defense Industries." *Journal of Business Management Perspective* 7 (25): 179–218.

Viotti, E. B. 2002. "National Learning Systems: A New Approach on Technological Change in Late Industrializing Economies and Evidences from the Cases of Brazil and South Korea." *Technological Forecasting and Social Change* 69 (7): 653–680.

Contributors

Mohammad Abbassi is Executive Manager, Sina Institute for Policy Making, Management and Innovation, Tehran, Iran.

Maryam Ashrafi is a PhD candidate in the Department of Industrial Engineering, Amirkabir University of Technology, Tehran, Iran.

Sanam Farnoodi is a PhD candidate in the Department of Management and Economy, Science & Research branch, Islamic Azad University, Tehran, Iran.

Mehdi Goodarzi holds a PhD in technology management andis a Senior Executive at the Ministry of Science, Research and Technology, Tehran, Iran.

Soroush Ghazinoori is Assistant Professor in the Faculty of New Science and Technology, University of Tehran, Iran.

Sepehr Ghazinoory is Associate Professor of S&T Policy, Tarbiat Modares University, Iran, and Editor-in-Chief, *Journal of Science & Technology Policy*.

Reza Jamali is a PhD candidate in Strategic Management, Tarbiat Modares University, Tehran, Iran.

Reza Mansouri is Professor of Physics, Sharif University of Technology, and Executive Manager, Iranian National Observatory (INO) Project, Tehran, Iran.

Manochehr Manteghi is Executive Director, Iran Aeronautical Organization, and former Executive Director of Iran Khodro Automobile Company, Tehran, Iran.

Tahereh Miremadi is Assistant Professor and a researcher at the Iranian Research Organization for Science and Technology (IROST), Tehran, Iran.

Marzieh Shaverdi is a Senior Specialist inManagement of Technology, Iran Atomic Energy Organization, Tehran, Iran.

Behzad Soltani, with postdoctoral research experience at Lund University in Sweden, is Assistant Professor of Mechanical Engineering, Kashan University, and formerly the Deputy Director, Iran Atomic Energy Organization.

Abdol S. Soofi is Professor of Economics at the University of Wisconsin-Platteville, Wisconsin, United States.

Parviz Tarikhi is a space science and technology specialist on microwave remote sensing, Iranian Space Agency, Tehran, Iran.

Fatemeh Salehi Yazdi is a PhD Candidate at the Manchester Business School, United Kingdom, and is a former expert on the Iran Nanotechnology Initiative Council (INIC), Tehran, Iran.

Mohammad Ali Bahreini Zarj is the Chief Executive Officer, Nanotech Fund, Tehran, Iran.

Name Index

Abbasid dynasty, 26
Abbassi, Mohammad, 12
Achaemenid Empire, 24
Aerospace Industries Organization (AIO), 194
Ahmadinejad, Mahmud, 12, 165, 181
Algeria, 233
Amirkabir University, 104, 106, 108, 204
Ansari, Anousheh, 211
Ashrafi, Maryam, 12
Asia Nano Forum (ANF), 134
Australia, 92, 207
Avicenna, 16, 28
Azerbaijan, 137, 233

Bahreini Zarj, Mohammad Ali, 11
Belarus, 233
Birooni, Abu-Rayhan, 15, 16, 28
Bouzjani, Abol-Vafa, 28
Brazil, 134
Bushehr, 162–164, 170, 174, 179

Canada, 97, 108, 165, 166
China, 29, 62–65, 99, 108, 134, 164–166, 187, 198–201, 207, 227, 236
Committee on the Peaceful Uses of Outer Space (COPUOS), 187

Daneshgar, 17
Daneshmand, 17
Darius, 24

Dar-ol-foonon, 18, 23, 31, 40, 41, 65
Deanvari, Abu Hanifeh, 28

East Asia, 2, 3
Egypt, 130, 140, 227
Eisenhower, Dwight, 161
Elm, 16–20
Ethiopia, 227
Expediency Discernment Council of the System, 67, 75, 93, 190, 213

Fajandi, Abu Mahmud, 28
Farabi, 27
Farahidi, Khalil Ben Ahmad, 26
Farnoodi, Sanam, 10
Farsi, 16, 20, 25, 27, 46, 75, 84, 93, 97, 98, 105, 109, 111, 189
Ferdowsi, 186
France, 45, 108, 159, 162–166, 168, 200

Germany, 45, 62, 108, 159, 162, 164–166, 180, 237
Ghazali, Muhammad, 30, 31
Ghazinoori, Soroush, 9
Ghazinoory, Sepehr, 10
Gilani, Koushyar, 28
Global Entrepreneurship Monitor (GEM), 59, 60
Godarzi, Mehdi, 9

Halal Internet, 103
Hong Kong, 3, 236
Hozeh elmiyeh, 18–20, 23, 34–35, 40, 143

Imam Hussein University, 204
India, 27, 134, 159, 162, 165, 166, 187, 200, 207
Indonesia, 207
Industrial Development and Renovation Organization of Iran (IDRO), 225
International Atomic Energy Agency (IAEA), 161, 182
Iran Khodro, 42, 219, 220, 225–227, 230–234, 236, 240, 242, 249
Iran Radiatorsazi Company, 220
Islamic Azad University, 43, 44, 77, 168
Islamic Development Bank (IDB), 130
Italy, 162, 187, 219, 236

Jamali, Reza, 10
Japan, 10, 78–84, 89, 102, 108, 123, 159, 164, 200, 207, 219
Jondishapour University, 21, 24, 25, 27, 40, 65

Karavan, 231
Kayhan Shenakht, 15
Khajeh Nasir Tusi Technical University, 204
Khajeh Nezam-ol-Mulk, 29, 30
Khamenei, Ali, 12, 50, 77, 93, 143, 181
Kharazmi, 28
Khatami, Mohammad, 117
Khayam, 28
Khomeini, Ayatollah, 42, 43, 111L

Malaysia, 130
Malik Ashtar Technical University, 204
Mansouri, Reza, 9
Manteghi, Manochehr, 13
Marvazi, 15
Mesbah Satellite, 193, 197–199, 208, 211–213
Middle East, 47, 62, 100, 108, 140, 170, 178, 191, 211
Ministry of Science, Research and Technology (MSRT), 7, 46, 51, 66, 69, 70, 72, 78, 193, 198, 201

Miremadi, Tahereh, 11
Mobaleghi, Hojatollah, 143
Mobile Communication Company of Iran (MCI), 90

Najaf, 15
Natanz, 164, 165, 177
National Research Council, 33, 36
Navid Satellite, 201, 213
Nezamiyeh, 19, 21–24, 29, 36–37
Nobakht, 27
Non-Aligned Movement (NAM), 134
North Korea, 187, 201

Office of Technology Cooperation, 48
Omavids, 26
Omid Satellite, 200–202, 208, 211, 213
Organization for Economic Cooperation and Development (OECD), 59, 127

Pakistan, 137, 198, 207
Papagan, Ardeshir, 21, 22, 40, 65
Pars Khodro, 219, 225, 226
Pasteur Institution, 139
Payam Noor, 77
Paykan, 219, 220
Persia, 9, 16, 17, 19–21, 24–26, 31, 36, 105, 190
Persian Gulf, 97, 202, 245
Pride, 231

Qajar, 22, 23, 31, 39–41
Qom, 16, 30, 37, 148, 150, 151

Rasad Satellite, 200
Razi Institute, 139
Razi, 16
Royan Institute, 11, 12, 143–153
Rubin, Uzi, 205
Russia, 45, 65, 89, 123, 159, 164–166, 196–202, 209, 211, 233, 236

Safir, 200–202, 206, 211
Saipa, 219, 225–227, 230, 231, 233, 235, 236
Sais Medical College, 24
Samand, 230
Sassanid Empire, 21, 24–26, 28, 36, 40, 65
Science and Industry University of Iran, 201, 205
Seda va Sima, 92
Senegal, 233
Shah Muhammad Reza Pahlavi, 186
Shah Reza Pahlavi, 186, 220
Shahab Ballistic Missile, 194, 205, 206
Shahid Beheshti University, 205
Shahid Sattari University, 204
Shariati Hospital, 147, 152
Sharif University of Technology, 32, 204
Shaverdi, Marzieh, 12
Shiraz University, 32, 145, 161
Sibawyah, 25, 26
Simorgh, 211
Singapore, 3, 139
Soltani, Behzad, 12
Soofi, Abdol, 10
Soofi, Abdulrahim, 15, 16
Soofi, Abdulrahmaun, 28
South Korea, 3, 4, 7, 8, 13, 123, 159, 198, 207, 218, 248
Soviet Union, 187
Spain, 164
Stanford Research Institute (SRI) Report, 161
Supreme Council of the Cultural Revolution, 44, 68, 79, 109, 143, 147, 165
Supreme Science, Research and Technology Council, 47, 68
Syria, 235

Tagh Kasra Arch, 21, 24
Taiwan 3, 4, 134
TAKFA, 94, 96
Tarbiat Modares University, 145, 147, 152, 205
Tarikhi, Parviz, 12
Technology Cooperation Office (TCO), 116
Telecommunication Company of Iran (TCI), 90
Thailand, 198, 207
Tunisia, 233
Turkey, 62–65, 102, 130, 137, 140, 207

Ukraine, 82, 134, 191, 233
United Arab Emirates, 97
United Nations Educational, Scientific and Cultural Organization (UNESCO), 91
United Nations Nuclear Non-Proliferation Treaty (NPT), 12, 161
United States, 42, 62, 71, 80, 88, 108, 117, 131, 159–162, 164–166, 172, 180, 186–188, 200
University of Tabriz, 41
University of Tehran, 19, 23, 41, 69, 73, 145, 149, 160, 161

Venezuela, 233, 235
Vietnam, 227

World Economic Forum, 60
World Trade Organization (WTO), 228, 242

Yazdi, Fatemeh, 11

Zohreh Satellite, 187, 188, 197, 208, 211, 212

Subject Index

Aerospace technology
　definition of, 185
Asian miracle, 3
Automobile
　assembly, 186, 219
　competitiveness, 44
　development in Iran, 218, 222
　imports, 218
　industrial policy, 218, 220
　industry, 223, 229
　industry twenty year vision, 237
　information technology, 103
　manufacturing in Iran, 218
　petroleum export, 218
　private investment, 221
　supercomputers, 106
　technological dependency, 222
　technology, 220

Benchmark, 11, 134
Brain Drain, 110, 246

Capital
　flight, 222
Capitalism
　crony, 222
　managerial, 80, 84
Catch up, 1, 117, 139
Chaebols, 8
Competitive advantage, 9, 122, 128, 237
Coordination failure
　government, 2–3
　market, 2–3

Cultural Revolution, 42–44, 55, 66, 75, 79, 90

Demand-side
　policies, 6, 7, 54
Developing economies
　problems, 3, 4
Development Plan
　first, 224
　third, 47, 220
　fourth, 46
　fifth, 46, 94–95, 172, 174, 208
　information technology, 94, 96
　management, 51
　nanotechnology, 134, 137
　nuclear technology, 174
　Office of Deputy President for Technical affairs, 69
　science and technology, 54, 75
　socio-economical, 45
Development stages
　efficiency driven, 60
　factor driven, 60
　innovation driven, 60
Developmental-state, 3
Digital Divide, 95

Economies of scale, 4, 8, 60, 80, 102, 111, 227, 240
E-Government, 100
Emerging economies, 1, 78
Emerging technologies, 45, 48, 49, 58, 67, 75, 76, 78, 115, 116, 118, 127, 160

Enterprise
    state-owned (SOE), 5, 7, 9, 45, 72, 77, 81, 82, 92, 226, 238
Entrepreneurial Activity, 59, 60, 63, 78, 141, 152
Entrepreneurship, 11, 59, 60–63, 83, 136, 147–149, 151, 153, 171
    definition of, 59
Era of decline, 30
Era of ignorance, 30, 32
E-Strategy, 89, 93, 94
    definition of, 94
Exports
    growth, 242
    Iran khodro, 231
    petroleum, 53, 55, 218
    software, 97

Filtering
    the Internet, 10, 104, 109–110
Firms
    followers, 4, 6, 7
    latecomers, 4–7, 186, 214, 230
    leaders, 4, 6
    start ups, 118, 121, 126, 140, 145, 151
Foreign direct investment (FDI), 1, 7, 237, 239
Foreign exchange, 218
    earnings, 223
    industrial enterprises, 238
    reserves, 230
    subsidy, 222
Foreign investment
    attraction, 167
    laws, 241
    sanctions, 110
Fuel
    efficiency, 229, 240
    subsidy, 104–105
Full Knocked Down kit, 242

General National Scientific Plan, 75, 185

Globalization
    Iran-khodro, 234
    Iranian auto industry, 229
    stages of, 219
Government policy
    coordination, 8
    remedy market failures, 7
Gross Domestic Product (GDP), 43, 45, 47, 59, 63, 70, 71, 77, 86, 102, 108

High-tech, 49, 76, 129, 169, 194
Horizontal policies, 122, 130

Imports
    illegal, 99
    law, 221, 223
    telecommunication equipment, 99
Import-substitution, 6, 7, 8, 220, 226, 240
Incubator, 48, 72, 121, 124, 125, 131
Industrial development
    definition of, 5
Industrial policy
    definition of, 5–6
Industrial transformation
    role of market, 6
    role of state, 6
Industry
    automobile in Iran, 220, 226, 236, 240
    Western, 218
Inflation, 76, 221, 240
    price of petroleum, 221
Information Technology (IT), 10, 34, 35, 74, 90, 93, 94, 96, 97, 102, 193, 194, 213, 235
Innovation
    definition of, 59
Innovation system
    national (NIS), 4, 39, 48, 55, 57–60, 62, 63, 65, 72, 74, 79, 82, 83, 127, 136, 241, 242
    sectoral (SIS), 48, 57, 140, 167

technological (TIS), 11, 58, 139, 141, 142, 146, 152, 153
Intellectual property rights, 7, 10, 111, 129, 144, 239, 241
International
  conferences, 130, 168
  cooperation, 12, 119, 185, 189, 206–207, 234
  exchange, 47
  exhibitions, 123
  interactions, 116, 171
  markets, 219, 231, 233
  networks, 133
  organizations, 133
  pressure, 197
  radio broadcasting, 92
  relations, 45
  robotic competition, 110
  S&T professionals, 55
  sales network, 80
  sanctions, 1, 11–12, 181, 208, 246–247
  science index, 118
  standards, 233, 238
Internet
  abuse by foreign powers, 110
  access, 109
  bandwidth, 107
  broadband, 107–108
  censorship, 10
  companies, 95
  cost of service, 111
  e-government, 110
  filtering, 110
  halal, 103
  investment, 10
  mobile connection, 91
  national network, 108
  relief and rescue operations, 187
  restrictions of use, 109
  satellites, 197
  security, 103, 107–108
  service monopoly, 107
  speed, 102
  subscription, 107
  users, 107
Iran Nanotechnology Laboratory Network (INLN), 117
Iranian government
  aerospace technology, 13, 185, 209
  allied armies, 89
  cancellation of nuclear contract, 163
  development of nuclear technology, 164, 178
  development of technology, 74
  economic role, 3, 5, 96
  emerging technologies, 76
  establishing telecommunication companies, 90
  motor vehicle industry, 229, 232
  spending on education, 70
  spending on R&D, 77
  State owned enterprises, 226
  Telecommunication Research Center, 89
  transformation of industries, 222
  U.N. Industrial Development, 134
  Western educational model, 39
Islamic
  enlightenment, 27, 28
  golden age, 18, 24, 25, 27, 28, 31, 36
  revolution, 15, 16, 17, 21, 23, 32, 35, 42, 43, 44, 66, 90, 111, 167, 180, 245
Islamization of the universities, 34

Joint venture, 1, 99, 232, 233, 237

Know-how, 8, 195, 222, 232, 247
Knowledge-Based
  development, 95
  economy, 39, 50, 78
Knowledge Diffusion, 72–74, 142, 152

Law
  budget, 119
  faculty, 149
  formation of organization, 167

Law—*Continued*
  government contracts, 99
  import of cars, 224
  investment, 167
  protection and development, 222
  tariffs, 225
  technology, 171
Licensing, 1, 7, 8, 246

Market-enhancing view, 3
Market failure, 3, 6, 7, 84
Market formation, 12, 75, 76, 78, 142, 150
Market-friendly view, 3
Moslem
  scholars, 18

Nanotechnology
  definition of, 116
National system of innovation
  definition of, 57
  functions, 58, 59, 142
Newly industrial economies, 2–4
Niche, 139, 142, 153
Nuclear Technology, 12, 159–173, 176–181
  quality control, 177

Original design manufacturing (ODM), 5
Original equipment manufacturing (OEM), 5
Own brand manufacturing (OBM), 5, 230

Petroleum
  crude, 159
  exports, 53, 218
  industry, 103
  IT application in, 106
  opportunity cost, 180
  price, 223–224
  research institute, 49
  reserve, 163

Policy documents, 50, 152
Policy instruments
  direct, 7, 8
  indirect, 7, 8
Political, Economic, Social, and Technological (PEST) Analysis, 110
Privatization, 8, 44, 81, 99, 226, 238
Property rights, 7, 111, 239, 241
  intellectual, 10, 111, 129, 144
Purchasing Power Parity (PPP), 59

Quality
  control, 128, 178, 239
  education, 43
  human resources, 95
  life, 117
  policy, 54
  product improvement, 78
  safety, 126
  services of, 224
  technical expertise, 110

Research and development, 4, 43, 67, 116, 117, 123, 126, 127, 130, 141, 144, 151, 165, 166, 170, 193, 194, 196, 247
Resource mobilization, 76, 142, 149, 151, 153, 154, 171
Roadmap, 50, 55, 152

Sanctions, 1, 10–12, 49, 53, 83, 109–110, 135, 159, 163, 169, 170, 177, 178, 181, 188, 208, 223, 224, 246–247
Science
  harmful, 20
  meaning in Iran, 16–17
  policy, definition of, 5
  prevalent concept in Iran, 19
  utilitarian, 20
Science and Technology (S&T)
  indicators, 33, 47, 48, 71
  parks, 48, 72, 124, 175, 237

Scientific
   credibility, 18
   prediction, 18
Scientist
   definition of, 17
   Islamic, 17
Semi Knocked Down kit, 242
Small and Medium-Size Enterprise (SME), 8, 72, 171
Social capital, 148
Socioeconomic Development Plan, 46, 47, 69
Spin-offs, 8, 75, 118, 150, 190
State owned enterprises (SOEs), 5, 9, 77, 82, 92, 195, 238
Strategic
   alliance, 1, 143
   cooperation, 233
   industries, 8
   plan, 12, 51, 52, 117
Subcontracting, 1, 5
Supply-side
   actors, 142
   policies, 7, 50
System failure, 141

Technological development, 2, 4, 5, 6, 12, 13, 46, 49, 50, 53, 69, 71, 76, 77, 78, 83, 115, 129, 153, 176, 180, 185, 246, 247
   demand side, 5–7
   stages of, 8
   supply side, 5–7
Technology broker, 247
Technology Transfer, 7, 42, 45, 81, 163, 165, 192, 203, 228, 231
Total Early-Stage Entrepreneurial Activity (TEA), 63

Un-grown-up talents, 32
Urban
   air pollution, 188
   topographical database, 196

Venture Capital, 60, 76, 118, 124, 131, 132, 150
Vertical policies, 122, 130
Vision 1404, 46, 75, 171, 237–239

War
   cultural, 110
   energy shortage, 164
   Iran–Iraq, 19, 148, 150, 223–224
   private consumption, 44
   social instability, 43
Welfare
   aerospace industry, 210
   consumer, 217

The manufacturer's authorised representative in the EU is Springer Nature Customer Service Centre GmbH, Europaplatz 3, 69115 Heidelberg, Germany. If you have any concerns regarding our products, please contact ProductSafety@springernature.com

Printed and bound by CPI Group (UK) Ltd, Croydon, CR0 4YY

23/03/2026

02076449-0011